W9-CDX-435

Race for Tomorrow

Race for Tomorrow

Survival, Innovation and Profit on the Front Lines of the Climate Crisis

SIMON MUNDY

WILLIAM COLLINS

William Collins
An imprint of HarperCollins*Publishers*
1 London Bridge Street
London SE1 9GF

WilliamCollinsBooks.com

HarperCollins*Publishers*
1st Floor, Watermarque Building, Ringsend Road
Dublin 4, Ireland

First published in Great Britain in 2021 by William Collins

3

A catalogue record for this book is available from the British Library

ISBN 978-0-00-839429-5 (hardback)
ISBN 978-0-00-839430-1 (trade paperback)

Typeset in Minion Pro by
Palimpsest Book Production Ltd, Falkirk, Stirlingshire

Printed and Bound in the UK using 100% Renewable Electricity
at CPI Group (UK) Ltd

MIX
Paper from
responsible sources
FSC™ C007454

This book is produced from independently certified FSC™ paper
to ensure responsible forest management.

For more information visit: www.harpercollins.co.uk/green

For the many people who contributed to this project in countless ways, and who are acknowledged more fully at the end of the book

Contents

Hellisheiði
Power Station
London
Dresden
Munich
Zurich
Venice
Haifa
Dhahran
Riyadh
Kori
Addis Ababa
Mustahīl
Lagos
Kinshasa
Kolwezi
Tiksi
Hunters' camp
Batagay
Yakutsk
Churapcha
Ulaangom
Ulaanbaatar
Jiangyin
Shanghai
Tsho Rolpa
New Delhi
Dhaka
Mongla
Jalna
Guangzhou
Shenzhen
Manila
Tacloban
Fulhadhoo
Malé
Indian
Ocean
South
Atlantic
Ocean
Mackay
Brisba
Sydney
Melbourne
Southern Ocean

Arctic Ocean
Qaanaaq
Greenland
Ilulissat
Nuuk
leistocene
ark
Cambridge
San Jose
Bermuda
LanzaTech plant
North
Atlantic
Ocean
Quidnet drill site
San Rafael del Norte
Pacific
Ocean
Fanalei
Boca do Acre
TI Uru-Eu-Wau-Wau
Diamantino
Toconao
São Paulo
Maule Valley
Puerto Ingeniero Ibáñez

Sometimes a story can seem so big

In the past few years, clim

from the margins to the

some of extreme weather

that this is some hypothe

where innovators seek wa

on every continent are fig

ties. With a snowballing

radical action, leaders fro

scrambling to position the

to build a new generation

energy and electric cars, as

this century.

But its scale and brea

Keeping up on it can mean

lots parts per million of ca

of deforestation, gigatonnes

energy investment. But its

of our present and future

taken from adaptation to tension

But that is not an option

century, you need to und

what is at present cascading thr

This book grew out of

and what I consider by

Preface

Sometimes a story can seem too big.

In the past few years, climate change has shot with stunning speed from the margins to the very centre of the global conversation. A surge of extreme weather events has crushed any lingering sense that this is some hypothetical challenge for unborn generations. While innovators seek ways to tackle the mounting hazards, people on every continent are fighting for the survival of entire communities. With a snowballing activist movement driving pressure for radical action, leaders from Washington to Brussels to Beijing are competing to position their economies for supremacy in a low-carbon world. A new generation of tycoons is chasing windfalls from clean power and electric cars, as the mighty fossil fuel industry lapses into existential crisis.

In its scale and breadth, the subject might look intimidating. Reading up on it can mean walking into a blizzard of abstract statistics: parts per million of carbon dioxide, thousands of square kilometres of deforestation, gigatonnes of melting ice, trillions of dollars of green energy investment. But it's now so entangled with every major element of our present and future, from what we eat to how we travel, from mass migration to tensions between superpowers, that skipping over this one is not an option. To understand the unfolding twenty-first century, you need to understand the climate crisis, and the changes that it's sent cascading through the modern world.

This book grew out of my own desire, as a journalist, to understand what I consider by far the biggest story of the century, and to help others to get to grips with it too. The toughest challenge, for anyone writing on this theme, is to avoid getting lost in the sprawling mass of information, with a subject that touches every part of the planet, every segment of the economy, and great swathes of science and technology. Yet beyond those statistics, one of the most compelling contests in history is under way, with a huge and diverse cast of characters from every walk of life and every corner of the globe. Whether they're fighting to make their fortunes or avoid disaster, they are engaged in a struggle that will shape the future for all of us. To tell that story, I realised, to capture the human drama behind the models and data points, I'd need to get out and meet the people on the front lines.

So I set out on a journey that ended up lasting nearly two years, through twenty-six countries on six continents. It took me to the edge of a fast-shrinking glacier high in the Himalaya, and deep into one of the hand-dug pits where Congolese miners are risking their lives to profit from the green tech revolution. I visited Amazonian tribes fighting to save their rainforest from illegal cattle farming, and an Israeli startup growing eco-friendly beef in bioreactors. I walked the shores of disappearing islands in the South Pacific, and through the frenetic clamour of China's biggest electric car factory.

Everywhere, I found people grappling with the unprecedented challenges that the climate crisis has thrust upon humanity. Some were at the base of the global wealth pyramid; some were billionaires. Climate change was opening tantalising new opportunities for some I encountered, while threatening to destroy the livelihoods of others. I met people rushing to build defences against catastrophe, and others jostling for leadership in the technologies that will power a transformed world economy. All of them are embroiled in a race that will set the course of our civilisation, and of the planet that houses it. These are their stories.

PART ONE

Thaw

CHAPTER 1

Thawing of permafrost is destabilising soils, human infrastructure, and Arctic coasts, and has the potential to release vast quantities of methane and carbon dioxide into the atmosphere . . .

– Intergovernmental Panel on Climate Change, *Special Report on the Ocean and Cryosphere in a Changing Climate* (2019)

YAKUTIA, RUSSIAN FEDERATION

Ten, perhaps twenty or fifty thousand years ago, these mammoth bones bore the weight of some of the largest mammals ever to walk the earth. Now they lie in odd little piles, jumbled with the skeletons of horses and bison that roamed with them on the ice age steppe of Siberia.

I crouch to pick through some of the bones strewn across the floor of the giant hole in the ground that I've clambered into. On every side, high above me, walls of cold grey earth soar before curving to form a dangerous overhang. The walls are moving – crumbling and retreating loudly as the ice within them melts, exposing the stringy white roots of ancient plants. High up at the rim, spindly trees stand at curious angles, as though peering into the hungry void that will soon engulf them.

Local people, steeped in the animist traditions of northeastern Siberia, have dubbed this place the gateway to the underworld. Scientists know it as the Batagaika Megaslump, the biggest phenomenon of its kind in the world. It began to form half a century ago in an expanse of coarse shrubs and larch trees, on a section where a makeshift road had been cleared. The disturbance caused underground ice to melt, creating at first a modest, barely perceptible dip. Its growth was turbo-charged from the 1980s onward as global warming gathered pace, the rising temperatures eating their way through new layers of frozen ground each summer. By the time I enter the megaslump, it's broad enough to hold 175 London buses laid end to end, deep enough to swallow the Sydney Opera House, and showing no sign of halting its expansion.

Vast as it looks, Batagaika is just a snip of a Russian permafrost zone that is the size of China, Afghanistan and Nigeria combined, and which has emerged as a potentially huge risk to the global climate. Beneath Siberia's frozen soil lies billions of tonnes of organic matter, the remains of ice age plants and animals. As warming temperatures thaw the permafrost, microbes are feasting on this material, releasing both carbon dioxide and methane, a still more potent greenhouse gas. Already, the carbon emissions from the Arctic permafrost are at a similar level to those from all international passenger flights. And even if all humanity were to stop burning fossil fuels overnight, the gases already in the atmosphere would ensure the permafrost kept thawing, releasing still more gas – a hideous cycle that could keep feeding on itself for decades or centuries.

I turn to find my guide Erel gazing at one of the ugly mounds that ripple along the crater's base, amid fast-flowing rivulets formed by melting ice. 'Over there,' he says, 'is where I found my tusk.'

The thawing permafrost might be a problem for the planet, but the emergence of mammoth remains has been lucrative for some in the semi-autonomous Russian republic of Yakutia, a place that brings home Siberia's staggering size. Yakutia's capital, Yakutsk, is 3,000 miles east of Moscow – putting it as far as northern Ethiopia from the Russian capital. Roughly the size of India, it has just 900,000 inhabitants who brave its absurdly harsh winters – with lows of -64°C in Yakutsk and -68°C in Oymyakon, respectively the world's coldest city and village. People with eyelashes frosted white shop at outdoor winter markets where fish, frozen rigid, are arranged upright like baguettes.

Yet Yakutia is now warming more quickly than almost anywhere on earth, with melting Arctic ice to its north creating an expanding blanket of dark seawater that absorbs ever more solar heat. The global average temperature rose by about 1°C in the past century, but parts of Yakutia are now warming at half a degree per decade. As the earth thaws and softens each summer, hundreds of Yakutian men head out to the wilderness in search of mammoth tusks. One of the prime sites is Batagaika, where those grim piles of bones have amassed as tusk hunters toss them aside.

Erel found his tusk in 2011 and sold it to a buyer from Yakutsk for $800, more than the average Russian monthly salary. But he was too early. In the years that followed, China cracked down on illegal elephant ivory from Africa – sparking a boom in Yakutian mammoth tusk hunting as prices soared. Engraved mammoth tusks can fetch more than $1 million on the Chinese market, after months of carving by artists in Hong Kong or Beijing. As they enter Macau's gaudy Grand Lisboa casino, high rollers are greeted by a series of intricately carved mammoth tusks: one showing the Chinese legend of the monkey king, another the Great Wall of China. Secretive traders, flying in from Yakutsk or China, take hefty cuts, but local hunters can still expect to make more than $10,000 for large, well-preserved tusks. Growing numbers of them have started going to serious lengths to beat the competition – leaving their families for months to camp in insect-infested hideouts, illegally putting the permafrost melt into overdrive by blasting at the earth with water pumps.

It's early evening when I jump into a small motorboat to travel 150 miles down the Adycha River. On both sides, trees are tilting and swooning where the river, swollen by melting ice, is fast eroding its banks. Around midnight the sun reaches its lowest point, still fully visible above the horizon but providing scant heat, and for five more hours I huddle shivering beneath a waterproof poncho.

At last I hear the crunch of the prow plunging into pebbles at the hunting site, hidden at a curve in the river amid hills dark with pine trees. A dog with thick white fur trots past rafts of yellow oil barrels, lashed together by the hunters to create platforms for their water pumps. Otherwise the site is deserted but for platoons of the mosquitoes that sweep the region in their billions each summer, in swarms so dense that I occasionally have to clear one from the back of my throat.

We notice an inflatable dinghy on the opposing shore, and cross to find a slim young man in a threadbare sky-blue jumper with cheeks dogged by acne. Aged 26, he's an out-of-work biology graduate here for a few weeks with his uncle in the hope of fast money. Last season he made a lucrative find at Batagaika. In a picture on his Chinese smartphone, he hunches with the muddy tusk slung over his shoulders, thrilled and exhausted.

He takes me to meet a group camped out further down the river – four chubby shirtless men, playing cards and smoking in a wood-framed tent hidden behind a curtain of bushes. They're still on edge after fleeing a police bust two days before, but one soon dons camouflage overalls and announces a visit to the hunting ground. A rival gang of five appears on that side of the river and watches us in silence as we cross.

We walk along a small valley filled with long beige hosepipes stretching from the water pumps at the river, and with the bones that the hunters have found and discarded. There's no Chinese demand for those, even the monstrous piece of mammoth hip I struggle to lift, about half my size and weight. But the area has acquired a reputation for rich tusk pickings. In one recent year the tuskers here kept a collective log of all their finds. The season's haul amounted to 2 tonnes of mammoth tusks, worth more than $1 million.

Months after I leave Siberia, I will have word from the young hunter, recounting how he left the crowded hunting site to wander alone in the tundra for forty-five days, betting that the warming earth would yield tusks without the need for pumping – or for sharing the bounty. The solo mission would yield enough ivory for him to clear his mortgage and buy a snowmobile.

'You know the gold rush that happened in California?' he tells me now with a grin in his riverside tent. 'This is the tusk rush.'

* * *

A few days later, 600 miles to the south, I board a midnight ferry across the Lena River, on a journey from Yakutsk to the village of Churapcha. While the thawing earth spells rich pickings for the tusk hunters, for the villagers here it's becoming a nightmare.

Semyon Nikitin is awake to greet me when I arrive at his home at five in the morning after a long drive through rutted roads, his eyes flinty but humorous above a thin white moustache. Semyon built the house of pale varnished planks three years earlier on retiring from the civil service. He chose this spot for the views, he says – and there is, at first glance, a fine vista of rolling green leading down to the woods a couple of miles away. But then I see them: the bouncing little hills like an endless series of burial mounds, extending from the edge of the forest right up to Semyon's back yard.

It's these weird shapes that have brought me to Churapcha, having intrigued me since I first saw them in satellite images on the computer screen of a scientist in Yakutsk. The land around the village seemed to have been overwhelmed by a rash of pustulous boils, rounded shapes bulging from the earth. The scientist – Alexander Fedorov, a chatty veteran at the famous Melnikov Permafrost Institute – told me that rising temperatures have caused a fourfold expansion of these so-called thermokarst landscapes in Yakutia since the 1980s. They're formed by melting ice, which produces ever deeper dips in the ground, Alexander explained. Sections of earth with lower ice content hold firm for a while, creating the bumps – but ultimately they too succumb, creating a huge bowl-shaped depression.

Semyon insists that his home will stand firm. He takes me outside to show me the protection he's put in place. The house is built on truck tyres and wooden beams, to insulate it from the ground. But his neighbours have already started to drift away. Snejana Titova, a 37-year-old accountant at the village administration office, despaired of the growing sums she was spending to address the large furrows appearing on her land. Now she's moving back in with her parents, who own a flat in the centre of the village where the underground ice content is lower. Even there the future looks uncertain, she worries. A few years ago, a visiting team of scientists from Japan and South Korea told the villagers that the proliferating grooves and dips could eventually grow into a depression enveloping all of Churapcha.

I heard a similar prediction from Alexander, who has spent forty years studying his homeland's frozen ground. If climate change continues unabated, he said, about half the population of Yakutia could ultimately need to move away from ice-rich areas that will turn into uninhabitable wastelands. 'We're trying to get people used to the idea that this landscape will change, becoming a swamp,' he said. 'And they will have to move, abandoning everything they have.'

Over lunch at Semyon's home, I notice a large portrait of Stalin hanging high on the wall to my right. Semyon is head of the local Communist party; now 69, he proudly pulls out his old Soviet military uniform, festooned with medals. But when he gives me a tour of the village, Semyon includes a large monument to local people who were dispatched in 1942 to some of the furthest reaches of Siberia, in one of Stalin's insane forced migration drives. The exiles struggled in their harsh new habitats, and two thousand died, according to an inscription on the memorial's outer wall. Now, Semyon and his neighbours are in danger of being uprooted again – due this time not to the whims of a tyrant, but to the power of natural forces warped by the pollution of modern civilisation. But in the distant northeast of this huge territory, an eccentric scientist is waging a lonely struggle to prove that Siberia can be part of the solution.

* * *

Like an Old Testament prophet Sergey Zimov strides ahead, white-bearded, stabbing the ground rhythmically with a tall metal pole, a thick halo of mosquitoes hovering over him. I'm in Yakutia's Kolyma region, an Arctic zone to the far north of Japan – a place seen even by Siberians as forbiddingly harsh and remote, whose biggest claim to fame was its brutal Soviet gulag system. Now, however, the area is starting to attract notice for a very different sort of initiative. It's called Pleistocene Park, Sergey's extraordinary project to fight climate change by slowing down the thaw of Siberia's permafrost. He wants to turn this landscape back to ice age 'mammoth steppe', grassland populated by large mammals – perhaps one day including the mammoth itself.

Earlier in the day, as Sergey steered his boat upstream from his research station, I asked him what had first attracted him to this distant outpost. 'Looking for freedom!' he bellowed over the deep groan of his outboard motor, as flecks of ash from his cigarette blew onto me.

Sergey had risen quickly through the Soviet scientific system before being sent to Kolyma in 1984 to study the local ecology. Ordered to close the station and move out after the Soviet Union's demise in 1991, he refused and turned it into a private research site. He'd been intrigued by the region's abundance of bones from the Pleistocene – a 2.5-million-year period, including a succession of ice ages, that ended about 11,000 years ago. Studying the remains, he concluded that this area had once hosted a great density of mammoths, lions, wolves and other animals, looking like a chillier version of a Kenyan safari park – until humans arrived in the region and hunted them to oblivion.

As the 1990s progressed, Sergey became obsessed by a second startling discovery: Siberia's thawing permafrost would release far greater volumes of greenhouse gases than other scientists had realised. His findings were overlooked for years before being published in *Science*, the world's most prestigious scientific journal. They helped galvanise a wave of new research into what is now seen as a major climate threat.

I can see why Sergey might have had trouble gaining recognition from the academic establishment. He revels in his status as a maverick scientist, prowling the land around the station with his long grey ponytail tucked under a beret, like a shorter, paunchier Fidel Castro, a cigarette never far from his lips, swilling vodka with every meal.

At Pleistocene Park, Sergey is putting his hard-won credibility to the test. He wants to prove that grasslands, roamed and fertilised by animals, can protect Siberia's permafrost from rising temperatures far better than the forest that now pervades the region. He hopes that if this initial project can prove his theory, he could secure funding from wealthy, green-minded donors to restore the mammoth steppe ecosystem across vast stretches of Siberia.

One afternoon in Kolyma, Sergey's son Nikita leads me through a field of sharp tree stumps, thinking out loud about setting fire to them. Forests in tropical regions are a crucial asset in the struggle against climate change. In contrast, Nikita says, the problem with the dark larch woods in the Arctic is that they absorb too much solar heat, warming the permafrost beneath them. Lighter, more reflective grasslands stay cooler – especially when blanketed in winter snow. If animals are plentiful, that snow is trampled into a thin layer that allows the bitter cold to permeate from the air, chilling the permafrost enough to protect it against thaw when summer returns. And so the Zimovs have set about their unique form of environmentalism, using chainsaws and a full-scale armoured tank to destroy their section of Siberia's famous taiga. 'I'm not a hippie,' Nikita tells me accurately, his blue eyes sharp above a curved scar on his chin.

We enter the main section of the park over a wooden slat bridge. A bison eyeballs us over a muscular shoulder before treading heavily away from two Mongolian sheepdogs – deployed to keep the herbivores moving, in the absence of real predators like wolves and lions. The dogs scamper after the bison before breaking into a sprint to chase a reindeer. At a distance are two clusters of sheep and musk oxen, lounging quietly. Later we pass a group of white Yakutian horses, their peroxide blond manes falling untidily over their eyes.

Today there are about seventy large animals in the park's 50 square miles of fenced territory. Funding and logistical challenges have meant that number has grown far more slowly than the Zimovs had hoped, and the project is still well short of the scale needed to prove their hypothesis. During my stay, Nikita is struggling with bureaucratic hurdles around the delivery of twelve bison from Alaska, bought with $150,000 from an online crowdfunding drive.

The bison quest looks straightforward compared with that for the Zimovs' ultimate prize: a resurrected woolly mammoth, which could dramatically accelerate the return of the grassland, felling trees with contemptuous ease. At Harvard University, a team led by celebrated geneticist George Church is working to engineer a mammoth, by selectively altering parts of the Asian elephant genome. Church claims he could start populating Pleistocene Park with mammoths within a decade. He has competition from South Korea's Sooam Biotech, which has partnered with Yakutsk's main university to search for mammoth DNA sufficiently well preserved to be cloned.

For now, the Zimovs continue their quest to repopulate their remote stretch of Arctic terrain with as many big beasts as they can get. It might seem laughably ambitious to turn millions of square miles from forest to steppe, but, they argue, it would cost a small fraction of the hundreds of billions that world leaders have committed to fighting climate change.

On my last day in Kolyma, I stand in a hollow recently created by the melting of a massive ice wedge, as Sergey looms lecturing above me. He's convinced that he has found a crucial piece of the answer to the global climate dilemma. But he's made sure to build his home on an ice-free patch of stony ground that will hold firm as the permafrost thaws. If the world decides to ignore his discovery, he'll keep developing his scientific theories on a warming planet, safe in his sturdy refuge as the coastlines flood and the deserts grow – and as the Siberian permafrost unleashes its carbon bounty in a terrible vindication of his warnings. 'I don't afraid if permafrost will melt,' Sergey mutters from the top of the newly formed hill, as the mosquitoes throng between us. 'It's not my problem. It's your problem.'

CHAPTER 2

Glacier retreat and permafrost thaw are projected to decrease the stability of mountain slopes and increase the number and area of glacier lakes. Resulting landslides and floods, and cascading events, will also emerge where there is no record of previous events.

– Intergovernmental Panel on Climate Change, *Special Report on the Ocean and Cryosphere in a Changing Climate* (2019)

NA GAUN, NEPAL

After a four-day trek up mountain paths filled with evil little leeches and lavishly expansive cowpats, my head feels like it's being squeezed in a vice as I sit in Furdiki Sherpa's kitchen hut, 30 miles west of Mount Everest in the Nepali Himalaya. Furdiki herself has never been troubled by anything so feeble as altitude sickness. At 74, she's spent her entire life in the village of Na Gaun, 4,180 metres above sea level, making her a prime witness to the thing that has brought me here: the huge, hazardous growth of the nearby glacial lake of Tsho Rolpa.

The world's mountain glaciers have become one of the most conspicuous indicators of global warming, now shrinking at a bewildering pace of about 335 billion tonnes a year, according to recent research – more than 10,000 tonnes a second. The impact of all that melted ice on global sea levels is a frightening prospect for low-lying islands and coastal cities. But there are some equally stark – and in many cases much more acute – threats to the communities right by the source of the problem, living on the slopes below the glaciers.

Na Gaun is the stuff of Nepalese tourism brochures, a scattering of tiny homes with sloping roofs, towered over by mountain ridges thick with snow. Between the houses stretch dry-stone walls, built to restrain the cows and yaks that ignore them to saunter imperiously through the village lanes. While the mountains dominate the skyline, when I stand among Furdiki's half-dozen cows in the yard outside her home, it's something else that catches my eye: the stony rim of Tsho Rolpa, a heavy crescent resting amid the peaks to our east.

Tsho Rolpa is one of more than 5,000 lakes formed in the Himalaya by water from melting glaciers, with thousands more in the Andes. NASA data shows their growth has been supercharged by global warming, with the area of the world's glacial lakes growing by over 50 per cent since 1990, and scientists are warning of a rising danger of sudden outburst floods. All that stands between Tsho Rolpa's 85 billion litres of water and Furdiki's home – and the homes of her neighbours, and all the other settlements I trekked through on the way up – is a fragile natural dam of rubble, riddled with air pockets and lumps of ice in what one researcher described to me as a 'Swiss cheese kind of structure'.

Seated on a low stool by her stove, Furdiki wears a bright purple fleece over traditional dress – a coarse, white-trimmed kimono-like outfit with a silver-buckled belt. Her face is deeply creased by the mountain climate, and by a broad grin that explodes across her face at the slightest provocation. Years of biting winter cold have given her cheeks a permanent blush that glows under her turquoise eyes.

As a girl, Furdiki spent two months each monsoon season supervising her family's livestock in the pastures that then surrounded Tsho Rolpa, sleeping in huts alongside the other village youth. 'Back then the lake was so small – about the distance from here to that house,' she says, pointing through her window to a dwelling just a few minutes' walk away. 'We used to lead the animals back and forth across the glacier.' But each rainy season, Furdiki would return to find the pasture shrunken, the ice scaled back, the lake enlarged.

Now, Tsho Rolpa is well over two miles long, and even to get close to the glacier – as I am to discover – takes several hours of fairly full-on exertion. A few weeks before my visit, a British–Nepalese team of researchers visited the site and warned that the glacier feeding the lake was retreating at a pace of 60 metres a year, with the lake now covering the area of 148 football pitches. The researchers urged the construction of new monitoring systems, warning of 'catastrophic' flood impacts for up to 6,000 households in villages including Na Gaun.

The valley already has a system of sensors and sirens, designed to warn residents of an outburst – although it would give people in Na Gaun minutes, at most, to flee their homes before the water hit. And even after a $3 million project in 2000 to create a new channel flowing out of the lake, slightly lowering the water level, Tsho Rolpa has continued to expand. Yet Furdiki refuses to share the scientists' view of the lake as a fearsome liability for her community. She's watched Na Gaun's population shrivel to a fraction of its size in her youth, and nearly all those left are in old age or near it. All seven of her children have gone to seek higher incomes in the Nepalese capital Kathmandu, and rarely undertake the exhausting trek to see their parents.

The only thing now breathing life into this ageing community, Furdiki says, is its lake. The more Tsho Rolpa grows, the more it catches the attention of trekkers and researchers, who are sustaining two lodges in the village. If the flow of visitors continues to grow, Furdiki hopes it might give her children a reason to come back to their home village. It's a perplexing thought – foreign visitors coming to gawp at a huge natural hazard, rejuvenating a village that could ultimately be destroyed by the object of their curiosity. But having come all this way, I can hardly skip the big attraction.

The next morning my guide Laxmi and I set out for Tsho Rolpa, following the trail to a cliff above the lake's western edge, then climbing a steep incline in the hope of finding a way down towards the glacier on the other side. After an hour of fumbling through the clouds it feels like a welcome sign of life when, near the highest point of our climb, Laxmi points out the paw prints of a carnivorous snow leopard. Just as I bend to look at them, what sounds like a sudden burst of cannon fire booms up from below. It's the decomposing glacier, a large block of it falling into the lake, and as we pick our way downward, trying not to slip on the carpet of loose stone, that truncated ice body comes into view.

Behind the glacier's jagged face its interior is a bright bubble-gum blue, offset by the dusty grey of the debris that lies on its roof, carpeting a still vast expanse of ice that stretches far back onto the slope behind it. At its edge, it's fast disintegrating, with small pieces plopping apologetically into the lake, and larger chunks producing those explosive sounds that echo high in the surrounding hills. Having reached the water they spread and drift, each slowly melting and making its contribution to Tsho Rolpa's growth. It seems only a matter of time – perhaps decades, perhaps days – before this process reaches its natural conclusion, with a violent escape of water triggered by an avalanche or landslide, or simply by the gradual weakening of that rocky wall at the lake's western end.

Not that such dark prognostications would wash with Furdiki, whose relentless good cheer leaves no room for visions of destruction. The theory of 'hyperobjects', coined by philosopher Timothy Morton, holds that the forecast impact of global warming is something on a scale so enormous, so divorced from our lived experience of the world to date, that we humans just aren't really built to process it. I'm reminded of that concept when Furdiki contemplates the notion of this village, where she has spent her entire life, being suddenly swept away. 'I've been saying the same thing to all those scientists who came here,' she tells me with her enormous kindly smile. 'I've lived here over seventy years and that lake has never burst out. Not once!'

* * *

From Tsho Rolpa I don't have to travel far to find a community that has already been hit by an outburst flood from a similar glacial lake. (Not far by Nepalese standards: a linear distance of 29 miles, covered circuitously through a fifteen-hour downhill hike; then nine hours alongside live chickens on a local bus plying roads of deep mud that send it rocking hilariously close to its tipping point; and finally an almost therapeutic ninety minutes in a pickup truck.)

Officially Tatopani is a village, although that term jars with its weird topography. It's really a series of ramshackle buildings in sporadic clusters, built beside a clifftop road running along the west bank of the Bhote Koshi, a river that originates in the barren peaks of southwestern Tibet. The Bhote Koshi here forms an international border; immediately after leaving Tatopani to the north, the road takes a sharp right over a bridge and lands in Tibet, a territory controlled by China since 1951. This has been a trading post for as long as anyone can remember, and as the Chinese economy grew in recent decades, so Tatopani benefited from the surging flow of goods over the border. But on the night of 5 July 2016, what rolled in from Tibet was something horrifically different.

Karma Lama, a mother of three then aged 49, was in her kitchen washing plates after dinner when she heard a distant roar. The family had moved into this home a few months earlier; it was still a work in progress, built with the profits from the tea shop they had been running for about a decade. It was 8.30 in the evening and her husband had gone to bed, leaving Karma and her youngest daughter to handle the dishes.

She stepped into the street to see one of the village's larger buildings swaying from the vibrations beneath it. Like every other resident I meet, Karma's first thought was that this must be another earthquake, following one that had damaged many homes the previous year. She roused her husband, who hurriedly pulled on a T-shirt, and they began clambering up the sheer slope on the other side of the road, dragging their 9-year-old between them.

Peering down in the darkness from amid the trees on the hill above, Karma couldn't see what was happening to her home or those around it, but her ears gave her a sufficient idea. Behind her, the deep rumble had grown into a deafening rush of water accompanied by the violent crash of boulders hurled by the engorged river against its sides. The family began walking through the night to Karma's ancestral village of Listikot, safely ensconced in the hills above the river but nearly ten miles distant. When they returned days later, their home was gone, along with the ground it had stood on and all their possessions. In its place was thin air above a near-vertical muddy slope, which ran from the edge of the now fractured road to the river below, still clogged with the rubble of about twenty houses destroyed by the flood.

The family left for Kathmandu, and after seven months in the dusty capital had saved enough to return and rent a space to relaunch their tea shop, where I meet Karma, her tattooed hands covered in chapati flour, hoop rings dangling from her ears. Each time she leaves the shop, she can see, just up the road, the empty space that once contained the house where she'd expected to spend the rest of her life. Now the family rents a small room in a nearby settlement, and it will be years before they're able to think about another home of their own. 'That flood came,' Karma says, 'and stole our dreams.'

* * *

For Nepal, mountain disasters are a threat to vital infrastructure as well as human life. Unlike its heavily coal-reliant South Asian neighbours, its grid runs almost entirely on low-carbon hydropower, with 99 per cent of Nepalese electricity generation coming from the rivers that course through its mountains. But the Himalaya's melting glaciers mean that many of these power plants are now under increasing threat from outburst floods. And this one showed just how grave that danger could prove.

A few miles downstream from Tatopani, the river runs into a huge artificial mound of earth and rock, behind which yellow-helmeted workers are milling around a collection of massive concrete walls bedecked with Chinese characters. In the height and thickness of its walls, the Upper Bhote Koshi Hydroelectric Project resembles some formidable coastal fortress. Yet this plant, too, was wrecked by the flood, and three years on, its reconstruction is still under way.

'The wall here,' the company's chief executive Bikram Sthapit tells me, as we stand in the mud on the downstream side of the dam, 'was blown onto the other side of the river.' I stare at him for a few seconds before I realise he's serious. The wall in question was a mammoth hulk of concrete, as tall as a decent-sized apartment building. It ran parallel with the river on its eastern side, housing a basin where silt was removed from the water before it was sent through a two-mile pipeline to the electric turbine in a powerhouse lower in the valley. The flood had blown the wall apart and into the sky.

Bikram was in Kathmandu when the flood hit the plant, which was then undergoing repairs following the previous year's earthquake. The only people at the plant that evening were two guards in the squat brick powerhouse, who looked outside the building to see prefabricated cabins collapsing. They ran. 'We were very lucky that it happened at night,' Bikram says. 'If it had happened in the day there would have been fifty or sixty workers in the powerhouse. Someone would definitely have died.'

He arrived at the plant days later to find the powerhouse had been totally flooded, destroying millions of dollars' worth of brand-new electronics. The scene at the dam was worse. The desilting basin on its eastern edge was smashed to pieces. On the river's western bank, a white-water rapid had formed where the flood had destroyed a curved section of the dam. Only the dam's strongest central section had held – but it was now peppered with huge boulders, and the flow was rushing right over it, so clogged was the riverbed with debris thrown down by the flood. One of the rocks found at the dam was 17 metres long, the size of seven Holstein cows.

Now, at a cost of about $70 million – roughly the same, Bikram tells me, as building the whole thing from scratch – the plant's reconstruction by engineers from China's Sinohydro is a few months away from completion. On the western bank, where water had blown through the wall, a new diagonal section has been put in place, to channel water towards the centre of the dam where it's strongest. The wall around the desilting basin has been completely rebuilt to withstand higher pressure; at the powerhouse, a new protective barrier has been built, and windows bricked up.

I ask if all this will be enough to save the plant in the event of another flood, with the number of upstream glacial lakes having quintupled since work first started on this plant over twenty years ago. If the same event were to happen again, Bikram replies, the design changes made to the rebuilt plant mean the damage would be much less severe than last time. But in the event of a larger inundation, he concedes, no such guarantees can be made.

Which brings me to one of the most unsettling elements of the Bhote Koshi flood story. In the disaster's immediate aftermath, there was total confusion as to what lay behind it, with initial news reports blaming heavy rains. Bikram commissioned a study from a Chinese consultancy, which established the following year that the cause had been the outburst of a Tibetan glacial lake called Gongbatongshacuo, 12 miles northeast of the plant. When we first met at his office in Kathmandu, Bikram showed me the satellite photographs they provided. In the first, taken before the flood, Gongbatongshacuo showed as a green bullet amid the grey and white of the mountain slopes. In the second it was gone. The water body's sudden disappearance made for an arresting pair of images. But just as striking, to my eyes, was its modest size. A short way to its south loomed the better-known glacial lake Cirenmaco – twenty times larger. And Furdiki's Tsho Rolpa, in turn, is another four times the size of Cirenmaco.

For all the devastation it visited on Tatopani and Bikram's power plant, Gongbatongshacuo was less than two hundred metres long. By the standards of Himalayan glacial lakes – by the standards now, let alone after the melting of the years and decades to come – this one was tiny.

CHAPTER 3

The polar regions are losing ice, and their oceans are changing rapidly. The consequences of this polar transition extend to the whole planet, and are affecting people in multiple ways.

– Intergovernmental Panel on Climate Change, *Special Report on the Ocean and Cryosphere in a Changing Climate* (2019)

NEAR QAANAAQ, GREENLAND

Drifting on still water 750 nautical miles from the North Pole, glittering in Arctic sunlight as it will be for another two weeks until the first sunset of August, Gedion Kristiansen abruptly starts chattering giddily under his breath. As he cuts the boat's engine, his son Rasmus silently lifts a bolt-action rifle and stares through its telescopic sight at a spot on the water where my untrained eye sees no sign of movement. As the shot echoes back from the barren beige crags that line the fjord, a scarlet cloud blooms from the fat body of a seal now floating lifeless on the surface amid scattered fragments of ice.

Gedion's Inughuit ancestors arrived in Greenland around the start of the thirteenth century, crossing with teams of dogs over the thick sea ice that divided it from what are now the far northeastern islands of Canada. The Norse colony founded 200 years earlier by the fugitive murderer Erik the Red – who famously gave Greenland its wildly misleading name in an attempt to attract settlers – was still alive far to the south, and archaeological evidence shows there was contact between the two groups, peaceful or otherwise. But while the Norse settlements died out in the late Middle Ages, the Inughuit lived on, the northernmost civilisation on the planet, isolated even from the other Inuit peoples who lived in more temperate pockets of Greenland's coast.

On this northwestern shore, only a thin strip of land peeked out from the ice sheet that covers four-fifths of the world's largest island, reaching a vertical thickness of up to 2 miles. Even on those exposed patches, no crops would ever grow in the dry, rocky soil, untouched by sunlight for four months every winter. And heavy sea ice meant fishing with boats was impossible for most of the year. But out on that ice was a rich ecosystem of blubbery animals that would sustain the Inughuit for centuries – glossy-coated seals, snarling 10-foot-long polar bears, wrinkled walruses with tusks reaching their chests. The Inughuit pursued them on sleds drawn by teams of a dozen dogs, and killed them with iron-tipped harpoons, the metal drawn from a meteorite that had smashed into the area millennia before. Successive generations of Inughuit men passed down the skills and knowledge needed to survive in this environment. Perhaps most critical was an ability to read the ice – to distinguish between *hiku-pajannguaq* (dangerously thin) and *hikuliaq* (thick enough to travel on, but slippery), between *maneraq* (broad and flat) and *manilarraq* (uneven enough to hinder dog teams), to name just a few terms from the huge glacial lexicon they developed.

The Inughuit's seclusion ended in the nineteenth century when they began to receive visits from traders and explorers like the American Robert Peary, leader of the first team to reach the North Pole. Peary took six members of the tiny community back to New York where he charged 20,000 people 25 cents each to look at them, before handing them over to be studied by the American Museum of Natural History, where they lived in the basement until four of them died of tuberculosis. In the 1950s, the Inughuit were forced to abandon several settlements and their burial grounds so that the US Air Force could build a base for its nuclear bombers, ready to take the transpolar route to Russia in the event of a third world war. Through it all they hunted on, now armed with rifles but still wearing the polar bear skins that repelled the cold better than any imported garments, and still deploying the intimate knowledge of the ice that had sustained them for centuries.

Now in his fifties, it's Gedion's turn to pass that knowledge to his son. But more and more of his time hunting with 24-year-old Rasmus is now spent on their motorboat, during an open-water season that his forebears took as a brief interlude from ice hunting, but which is now taking up an ever larger chunk of the year. Since 1979, the

Arctic melt period has been lengthening by ten days each decade, falling into a self-perpetuating feedback cycle. The expanding area of dark seawater absorbs much more light than the reflective ice it has supplanted, driving more warming and yet more melting – a spiral that helps explain why the Arctic is warming at triple the global average rate. Even where the ice does still appear on the ocean it's now, on average, 40 per cent thinner than it was when Gedion was a teenager.

At first, the shift seemed to bring some benefits. In the 1990s, Inughuit hunters started bagging unprecedented numbers of polar bears, as the shrinking sea ice forced the animals to hunt on a smaller area of ice near the coast. The hunt became so fruitful that the government imposed kill quotas, fearful that the bears would be hunted to extinction. But as the change to the ice accelerated, it began to pose new hazards even to Gedion, one of his community's most respected hunters. One day he was hunting near the ice edge when he saw that the section he stood on had become detached from the rest of the sheet, leaving him stranded. Realising his mobile phone had signal, he called for help and a rescue helicopter was sent from the US air base. While he waited, he watched the ice floe breaking up around him. The chopper arrived and hovered overhead as the crew hauled Gedion and his dogs aboard. There was no room for the hefty sled; as he flew away to safety, Gedion looked down to see it had already vanished into the water.

An upbeat type, Gedion is unsentimental about the changes. To plan safe routes during the ice season, he notes, the hunters can now use satellite images posted by government officials in the centre of Qaanaaq, the village where most of the 1,200 Inughuit now live. And for all the mournful howling of his dogs, which now spend nearly half the year chained to rocks by a stream on the village outskirts, it's just an easier life moving around by motorboat, he says.

Other hunters in Qaanaaq sound a more sombre note. Jens Danielsen, the village's heavyset former mayor, had distrusted the environmental movement since the 1970s when the French actress Brigitte Bardot led a campaign that destroyed the international trade in sealskin, then the linchpin of the Inughuit economy. But in 2015, when the world's green activists descended on Paris for the annual United Nations climate conference, Jens and Gedion's brother Mamarut travelled there too, and made appearances in their hunting furs to raise awareness of the threat to their way of life.

'Everything is changing now,' Jens tells me one afternoon in the kitchen of his wood-framed home, brightly painted like most of Qaanaaq's houses, perhaps to temper the gloom of the sunless winters. On the front porch, his wife and children are slicing a section of narwhal that his son caught on a recent hunting trip, sneaking up on it in a kayak before spearing it with a harpoon. In the past, the hunters would always catch narwhals by paddling out from the edge of the ice; now, during this expanding melt season, they carry their kayaks out to sea by motorboat. 'I remember when I first started hunting,' Jens says. 'I was about fourteen. The sea would freeze at the start of October, when there was still some daylight, and we used to take the dogs up north, near Canada, to hunt polar bears. Now these are all just stories I can tell my grandchildren.'

Aged 61 and struggling with arthritis and heart problems, Jens is hunting less these days. And while many of Qaanaaq's younger men are just as enthusiastic hunters as their fathers, the rapid change in the conditions means that they will never achieve earlier generations' mastery of the winter sea, honed over centuries of observation and experience. 'The great hunters are getting older now,' says Aleqatsiaq Peary, the great-great-grandson of the famed American explorer and Aleqasina, an Inughuit teenager. 'They knew everything about the ice, how it behaves, the thickness and movement. They knew how to walk the ice just by looking at it. There won't be any great hunters in the future.'

* * *

A short distance south of Qaanaaq, less than a day's dog-sled ride, lies the abandoned settlement of Moriusaq. Once home to more than a hundred people, its population dwindled rapidly in the late twentieth century until just four hunters remained. After a gunfight broke out between them in 2009, the three survivors left the village for good.

When Rod McIllree's boat docked near Moriusaq's empty houses a few years later, he saw not an eerie ghost village but a career-defining opportunity. A geologist trained in the mining country of Western Australia, Rod had been drawn to Greenland by the chance to be a pioneer on a new frontier for the industry. Just on the fifth of the island that's not covered by ice, geologists have discovered scores of valuable mineral deposits, from rubies to uranium. The problem, particularly in these northern reaches of the island, was the sea ice. Historically, the waters around Moriusaq could be traversed by ships for only two months a year, Rod tells me when we meet in his office in London's Mayfair before my visit to Greenland. Now that period has more than doubled. And that's enough to open the way for a lucrative operation to extract what Rod and his team discovered on Moriusaq's strip of the Greenlandic coast: one of the world's highest-grade deposits of titanium ore.

Without the longer melt season, Rod says, a mine at Moriusaq would have been out of the question. Because the northern Arctic is still littered with ice floes even in summer, bulk carrier ships with specially reinforced hulls are needed to transport minerals – and because there are very few mines in the region, shipping companies have very few of those ice-class 'bulkers' available for hire. Rod's small startup, Bluejay Mining, could never have secured enough ships to carry out a whole year's mineral production from Moriusaq in just two months. But with four or five months, they'd have time to make return trips. The maths worked out: this project could generate a substantial return – big enough to secure the support of British financial giants like Prudential, HSBC and

Barclays, who pumped in the millions needed to start making the mine a reality.

Rod expects to start shipping titanium out from Moriusaq in 2023, and other exploration companies are jostling to follow with more mining projects on remote areas of Greenland's coastline. 'These environmental changes have brought a realisation that Greenland might not be too challenging,' Rod says. 'There are projects being contemplated now that would never have got off the drawing board ten or fifteen years ago. I mean, it's just accelerated out of the box.'

The impact of the Arctic's melting ice is reverberating far beyond Greenland, feeding into the complex tensions of twenty-first-century geopolitics. During my travels in Siberia I spent several days in Tiksi, an Arctic port built under Stalin's rule in the 1930s as part of plans to develop the region. After the collapse of Communism, traffic through Tiksi fell to almost nothing. In a market economy, there had to be a damn good reason to send expensive ice-class ships to such a remote location. The population plunged by more than 80 per cent. In the perpetual grey mist of a Tiksi summer, I wandered its streets of mostly abandoned buildings, their naked doorways giving views into floors strewn with debris. But I found Alexander Chusovsky, the port's director general, on bullish form. Vladimir Putin's government had authorised a huge expansion of the Tiksi port, more than quadrupling capacity. It was part of Putin's plan to exploit the Arctic ice melt to make Russia's northern waters a major shipping route between Asia and Europe, with a distance 40 per cent shorter than the route through the Suez Canal. Putin had given his officials seven years to achieve a tenfold increase in traffic. Soon after my trip to Tiksi, a vessel owned by Danish shipping group Maersk became the first container ship to complete the route, reaching Germany from South Korea in twenty-five days.

Russia's Arctic strategy – especially its assertion of national control over the nascent cargo route – has set alarm bells ringing in

Washington, according to Michael Murphy, the US deputy assistant secretary of state who handles the Arctic security portfolio. 'It is one thing for Russia to invest in its own Arctic littoral,' he tells me on a call after my stay in Greenland. 'It's another thing to be developing a capacity to project power into the North Atlantic and the Arctic in the way that occurred during the Cold War.'

Having joined the foreign service in 1991, the year the Cold War ended with the Soviet Union's collapse, Murphy is now in the Arctic theatre of a new age of tensions – not only with Russia, but also with China, which is ramping up investment in the region while proclaiming itself a 'near-Arctic power', despite the 1,200-mile stretch of Siberia that separates China's northernmost point from the nearest Arctic shore. When Chinese state companies nearly sealed a deal to build a new generation of airports in Greenland, it was a wake-up call for the US government, Murphy says, noting that New York is closer to Greenland than it is to California. And with Arctic shipping potentially set for a meaningful role in the global economy, the region has 'risen dramatically' as a priority for the US State Department.

'Fifteen years ago people wouldn't have posited that the circumstances we have today, or the trend lines we're on today, were even possible,' Murphy says. 'What we can't do is wake up one day and find all this worked against us and our long-term interests.' In Donald Trump's administration, the strategising around Greenland went all the way to the president, who publicly mulled a bid to buy the territory from Denmark, which has controlled it since 1814. 'Essentially it's a large real estate deal,' Trump told reporters. He then cancelled a state visit to Copenhagen after his suggestion was dismissed by Denmark's prime minister 'in a very nasty, very sarcastic way'.

'We thought it was a joke, at first,' Greenland's mining minister Jens Frederik Nielsen tells me in his office in Nuuk, the capital city of 18,000 people that sits on the island's southwest coast. 'But it shows how interesting Greenland is to the world.' Nielsen is 29 and looks a few years younger, and while he seems well on top of his brief, his swift political ascent may owe something to what he warns is a key problem for his homeland: a shortage of modern professional skills, with most Greenlanders leaving high school early to start work in their teens. It's a headache for his government, with its long-term ambition of independence from Denmark.

But the melting ice is transforming the economic outlook, brightening the prospects for an independent Greenland. 'It will be much easier to sell our country and make it attractive' to more miners like Rod, as the shipping season lengthens, Nielsen says. Under a 2009 agreement with Copenhagen, Greenland's government has already secured the right to keep all the tax revenues from mining on its territory, putting it in line for a potential bonanza. As the decades pass and new Arctic routes open, the maritime distance to China's resource-hungry factories could be cut in half from the 13,000 miles that ships would need to travel today on a circuitous route through Suez. 'It's not just the mining industry – this will benefit every industry in Greenland,' Nielsen says. On his windowsill, a wooden carving of a hunter's kayak sits as a silent reminder of the old traditions that are being upended by the same process that's driving the mining growth. 'Culturally, the melting ice is a disaster,' the minister concedes. 'But from an industry point of view, it will be really, really helpful.'

With a population small enough to fit inside the West Ham United football stadium, his would-be nation has no choice but to make the best of the changes that global warming has forced upon it. But many millions all over the world will feel the impact of Greenland's melting ice – not just around its shoreline, but across the white desert that sprawls for 650,000 square miles over its frozen interior.

* * *

As we drift lower over the ice sheet, the sense of ethereal beauty gives way to something roughly opposite. From this shallow height the ice appears smashed and broken, like an incalculably vast slab of white marble after a protracted carpet-bombing campaign. What looks from greater distance like a gently rippling Christmas cake is revealed as a jagged landscape of spikes and shards, sliced by a network of crevasses running in all directions, veined by dark streaks of soot that have floated from chimneys and exhaust pipes across the northern hemisphere. Our tiny plane banks sharply (I've hitched a ride on a training flight for two locally based pilots) and follows the ice to the point where, between a pair of featureless grey hills, it collapses into a soup of white slush, floating on the surface of the fjord.

This is the Jakobshavn Glacier, the most famous segment of the ice sheet that covers most of Greenland. Each year, it deposits tens of billions of tonnes of ice into the fjord at its edge, with blocks dozens of metres high amid a mess of smaller fragments. At the fjord's mouth, icebergs the size of pyramids drift into the dark ocean, saluted by the waving tailfins of frolicking beluga whales and small boatloads of tourists photographing one of the world's great vistas.

But while the icebergs are the main draw for most foreign visitors, it was those crevasses that caught the attention of the Swiss scientist Konrad 'Koni' Steffen, on one of his annual flights over Jakobshavn. 'It looked like someone had taken a big sledgehammer and just started hitting the glacier,' he tells me in the lounge of his hotel in Ilulissat, the small fishing town that adjoins the icefjord. 'It was the acceleration, obviously. As the ice moves faster, it pulls apart.'

The glacier had always shed ice into the sea, while regaining mass at its other end from snowfall far inland. It's widely seen as a likely source of the iceberg that sank the *Titanic*. But from the late 1990s, the glacier began losing mass at an accelerating pace, becoming the most conspicuous site of the shrinkage of Greenland's ice sheet. And that shrinkage was happening far more quickly than previously thought – something that we know, in large part, thanks to Koni Steffen, one of the most influential and charismatic scientists to have turned their attention to Greenland's frozen wasteland.

With his full grey mane and thick beard, Koni is a throwback to an earlier, buccaneering generation of polar scientists, for whom research meant disappearing into the ice sheet for months on end. His intrepid reputation dates back to the 1970s when, as a graduate student on field research in Canada, he was hit by an avalanche that left him with a dislocated jaw, a shattered leg and no idea who he was. Having rediscovered his identity by reading through his field notes, Koni penned a farewell love letter to his girlfriend Regula. Rescued after ten hours, he went on to marry and raise two children with her. But on future forays into the Arctic, he kept that letter safely stowed inside his jacket.

Far from being scared away from the ice, Koni went on to spend more time on it than almost any other scientist of his generation. In 1990 he secured funding from his Swiss university, ETH Zurich, for a project studying Greenland's ice sheet at the 'equilibrium point', where winter snowfall and summer melt are in balance. This was the birth of Swiss Camp, an institution that would become legendary over the next three decades, drawing visits from European royalty and US congressional leaders. It had a kitchen, sleeping quarters and a propane-fired sauna, sitting on a wooden platform that rested on the surface of the ice, buttressed by poles driven deep beneath the surface.

Every year Koni returned for weeks or months at a time with a small team of graduate students. Those picked to join him could expect a gruelling routine, with long journeys on Ski-Doo snow-mobiles to maintain and upgrade the measurement instruments that Koni had installed across this region of the ice sheet. In the evenings, the researchers would assemble at a round table for dinner, the conversation flowing long into the polar night. 'Everyone has their turn to cook and we're very strict: it has to be a three-course dinner,' Koni says. 'If you work fourteen hours outside, you're hungry.'

As the years progressed, Koni realised that his work on the ice sheet was uncovering something that would be felt far beyond the world of glaciology. During the first two decades of his work at Swiss Camp, the average winter temperature there increased by 7.3°C. As the melt accelerated beneath it, Swiss Camp's foundation poles, once penetrating several metres into the ice, now held the platform high above the ground like a house on stilts. One spring Koni returned to find the whole thing had collapsed. He rebuilt it, only for the same thing to happen again a few years later.

And it wasn't just on the surface that the ice sheet was changing. From his GPS readings, Koni realised the ice beneath Swiss Camp was sliding towards the sea far more rapidly than should have been expected. As the ice moved faster, it was stretching and splitting apart, opening deep cracks in the surface. 'Thirty years ago there was not a single crevasse in the area,' he says. 'Now, every fifty metres we have a crevasse big enough to put the whole Ski-Doo in there. It's no longer safe to walk around.'

Greenland's ice sheet has always been in motion, melting away at its edge and replenished by snowfall in the middle. But an acceleration of this order sat totally outside the established science. Koni's own professors had taught him that ice sheets responded to climatic changes over centuries or millennia. Now he realised that the glaciologists who came before him had got it wrong. Analysing the data from his measurement stations, in collaboration with NASA scientist Jay Zwally, Koni figured out what was happening. Through cracks and fissures in the ice, meltwater from the sheet's surface was trickling down hundreds of metres to the bedrock, where it acted as a lubricant, making the ice slip more quickly away from the centre of the island. And the faster the ice moved, the more cracks opened within it, allowing more water to reach the bedrock and make the ice move even more quickly.

The faster-than-expected shrinkage of this ice sheet – and of the even larger one in West Antarctica – has already begun to drive what will be the most rapid sea level rise for many thousands of years. According to a 2019 report from the Intergovernmental Panel on Climate Change, the global average sea level is likely to rise this century by between 26 and 110 centimetres. Anything within that range will have serious consequences for coastal populations. And Koni – a lead author of that IPCC report – says that the reality could be far worse. Any report from the IPCC is 'very conservative', he says. 'Otherwise it would never be authorised by all the governments. But the inside knowledge is, it could be one metre, it could be two and a half metres, to be honest. The one metre is almost guaranteed, even if we stopped all CO_2 emissions today.'

For all the grim import of his work, Koni doesn't try to hide the thrill it gives him as his career draws to a close. Next year, with the compulsory retirement age of 70 looming, and with no other researcher willing to take responsibility for Swiss Camp, Koni will destroy it as required by Greenlandic environmental law: sending valuable equipment out by helicopter and burning the rest (a smaller carbon footprint than extra chopper trips, he explains). Tomorrow, he'll make a penultimate visit to the site – this time for just a week, after the coronavirus pandemic forced him to cancel what would have been his final spring visit with his students. It's now August, peak melt season, making the ice much more treacherous than in the spring. Adding to the challenge, an unusually intense snow storm has just hit the area, throwing a cloak over the crevasses that have proliferated around Swiss Camp. And, yet again, the camp has collapsed, meaning that Koni and his three companions – including his climatologist son Simon – will need to build a temporary shelter when they land. 'I'm actually glad it collapsed,' Koni says. 'Now I know it's gone. Now we can move it out.'

But even as he prepares to draw down the curtain on Swiss Camp, Koni is buzzing at the prospect of getting back on the ice. 'Tomorrow will be pure survival,' he says. 'We'll have to break into the platform that collapsed, cut through to the boxes with our tents inside. But within two days we'll have a new camp erected on the site. And then' – his eyes sparkle – 'then we start the *science* again.'

Three days after arriving at Swiss Camp, Koni left Simon and the other scientists working on the ice to walk to a measuring station a short distance away. The heavy snowfall had resumed and they quickly lost sight of him; when he didn't come back they assumed he had gone to his tent to rest. But on their return to the camp after several hours, there was no sign of him. They retraced his steps, which terminated at one of the treacherous crevasses that Koni had described to me, carved into the ice by the force of global warming. At its bottom was a newly formed hole, two metres long, leading into a water chamber within the ice sheet. They knew immediately that there would be no chance of finding him alive.

Like few other scientists of his generation, Koni Steffen expanded our understanding of the profound impact that climate change will have on our planet, of the silent changes in its polar reaches that will ripple with devastating force through every region of the globe. The oceans, we now know, have already begun to rise with a speed that would have been unimaginable a generation ago. The question now, for people in low-lying places all over the world, is what to do about it.

PART TWO

Rising Tides

CHAPTER 4

In general, it is technologically feasible and economically efficient to protect large parts of cities against 21st century sea level rise. However, questions of affordability remain for poorer and developing regions.

– Intergovernmental Panel on Climate Change, *Special Report on the Ocean and Cryosphere in a Changing Climate* (2019)

LAGOS, NIGERIA

Within a massive expanse of newly created land on the West African coast, Ronald Chagoury Jr stands next to me on the bridge, looking down at a canal lined with young trees planted at even intervals. 'We were inspired by the Seine, in Paris,' Ronald says. 'And this' – he gestures further down the road on which we stand, where the bridge leads into a huge multilane boulevard, shimmering in the midday heat – 'this was inspired by Fifth Avenue, by the Champs-Élysées. You can see the size of the road, of the sidewalks.' Indeed, their scale is hard to miss, like much else in this gigantic new development taking shape on the shoreline of Lagos, Nigeria's economic hub and Africa's largest city, with roughly – no one can be sure – twenty million inhabitants.

For years, the ocean ate away at Victoria Island, Lagos's most upmarket area and home to its most important commercial offices – a threat that became more severe as climate change took hold, with the sea becoming higher and more powerful. Today the streets of VI, as it's universally known, are more clogged than ever; moto taxi riders, jaws locked in concentration, pick their way past bankers in stationary black Land Cruisers, as hawkers squeeze in their dozens, bearing refreshments and pirated books, through the gaps between the static vehicles. But the menace of erosion has receded. Next to the traffic jams of Ahmadu Bello Way, where once stretched the rapidly disappearing Bar Beach, now lies 6.5 million square metres – and counting – of land reclaimed from the ocean and raised high above it, enormous quantities of sand protected by a tall barrier of boulders and cement dubbed the Great Wall of Lagos. This is Eko Atlantic, a climate resilience project that has few rivals in its size and audacity, pursued by one of Africa's most wealthy – and controversial – business families.

With its huge empty avenues Eko Atlantic today reminds me less of Paris or New York than of Naypyidaw, the grand, sparsely populated new capital city thrown up by Myanmar's military rulers. But there is a serious business case for it, insists Ronald, scion of the billionaire Chagoury clan and vice-chairman of South Energyx Nigeria, the family company building the project. The fearsome overcrowding in other parts of the city is in itself a decent incentive for developers to build homes and offices in this new district, with its broad thoroughfares arranged in an orderly grid. And while Eko Atlantic and its wall may have fended off the erosion from ocean waves, VI and much of the rest of Lagos still face a growing threat from flooding. Downpours during the mid-year wet season, which drove lethal inundations in the months before my visit, are set to become increasingly severe. As the sea rises, so too must the lagoon around which Lagos is built and from which it takes its name – further weakening the effectiveness of the clogged, antiquated drains that are already doing a shoddy job of keeping the city's streets and buildings clear of water. Eko Atlantic, with its state-of-the-art drainage system leading straight into the ocean, will only look more attractive the more the problems mount in other parts of Lagos.

Building this sturdy enclave from the sea has been a long, stressful and expensive saga, Ronald says. He's been working on the project since returning in 2005 from engineering studies in Paris. Two years earlier the idea had sprung from a conversation between his father and Bola Tinubu, then the governor of Lagos state and now national leader of Nigeria's ruling party, who was concerned about the accelerating shrinkage of VI. Ronald was taken aback by the scale of the plan – 'I said, this project is crazy. It's stupid. It's gonna take our lives to build this thing' – but soon became enthused, travelling to a cavernous wave-testing facility in Denmark where engineers lashed a prototype section of the Great Wall with once-in-thousand-year storms. It passed that test – although the family later lost more than $10 million in a single night when a real-life storm devastated an unfinished section of the structure.

Crowned with a paved promenade, the wall now runs for two kilometres and will ultimately reach three, making it the third-largest structure of its kind in the world, Ronald says. On the expanse of sand behind it, there are just a handful of completed towers and a few more under construction, all of imposing height but far outnumbered by the empty plots with their yawning foundation pits. Ronald is adamant that the pace of development is no cause for concern. It's only four years since the first building went up, he says. With a target resident population of 350,000 and a further 250,000 commuters, this will be virtually an entire new city tacked onto the existing one – and cities take time to build.

In Ronald's telling, his family has created something that will safeguard the future of Lagos, enabling the city to weather the challenges of climate change and achieve a clout to match its soaring population. As we look out over the barren landscape, he paints it in the colours he promises it will one day assume – 'a financial hub, a tech hub, a touristic centre'. Together with the country's buzzing music, film and fashion scenes, Ronald insists, the Chagourys' colossal venture can help to change the global image of Nigeria, and of all Africa.

The Chagoury clan's own image could also use a boost from the project. The family arrived in Nigeria as part of a small but successfully entrepreneurial wave of migration from Lebanon into West Africa that began in the late 1800s when, according to legend, a few Lebanese merchants disembarked by mistake en route to South America. Gilbert and Ronald Sr – respectively Ronald Jr's uncle and father – went into flour milling, and then construction. Their star reached its zenith when they became close associates of the 1990s military ruler Sani Abacha, generally regarded as the most deeply kleptocratic of the leaders who have ruled Nigeria since independence, estimated to have funnelled about $4 billion of public funds into his personal coffers. The full details of the Chagoury brothers' operations during those years remain unclear, but they landed Gilbert with a money-laundering conviction after a Swiss investigation into Abacha's financial manoeuvres. This background, and the Chagourys' continuing connections to leading politicians in a country infamous for graft, has led to mutterings about the true origins of the funding for Eko Atlantic. 'They are thugs,' was the brutal assessment of one member of Lagos's business world over drinks in a quiet corner of a VI bar.

It would be hard to apply that epithet to the urbane, vegan 36-year-old Ronald Jr, who has embraced the millennial health fad of regular day-long fasts, and since yesterday's lunch has been consuming nothing but mushroom coffee. With his African birth, Middle Eastern heritage, French schooling and mid-Atlantic twang, Ronald is almost a caricature of today's globally oriented, ecologically minded young executive class as he lays out his hopes of making Eko Atlantic a green city fit for the twenty-first century, powered by solar panels and roamed by emission-free buses.

However green it might be, Eko Atlantic won't be cheap, at least not by the standards of a country where gross national income works out at about $2,000 per person. That has brought some biting assessments of the entire project as an enclave for the elite, who will stay dry in their sheltered towers while poorer citizens wallow in rising oceans. A 2017 report in *Rolling Stone* magazine dubbed it 'climate apartheid', alleging: 'Even more than other gated communities, Eko Atlantic says to the world, No, we are not all in this together.' Ronald raises this attack unprompted, remarking caustically that such criticism of this homegrown African project has come overwhelmingly from journalists in rich nations like the United States that account for the bulk of the emissions driving climate change. The first residential towers appearing here are for the luxury market, he concedes, but in time there will be flats available for as little as $200,000, aimed not at oil millionaires but at the rising professionals of the new Nigerian economy. Still, I quickly calculate, the average Nigerian – even if she could put all her income towards the purchase – would need to work for a century to earn that sum.

One person who has been impressed by the Chagourys' vision is Bill Clinton. The former US president made an effusive congratulatory speech at a 2013 dedication ceremony for Eko Atlantic, which is among the 'global change' projects featured on his charitable foundation's website. The relationship caused problems for his wife's 2016 presidential campaign, as media homed in on Gilbert Chagoury's multimillion-dollar donations to the Clinton Foundation and the subsequent decision – under Hillary Clinton's authority as secretary of state – to look into building a new US consulate at Eko Atlantic.

When I visit, Donald Trump is still in the White House; given his open loathing for the Clintons, I wonder whether the consulate plan has now been shelved. On the contrary, Ronald says, the deal was signed just a few months ago. A construction project of prodigious scale and expense, liberally sprinkled with luxurious penthouses and gleaming office towers – perhaps this is the kind of climate resilience initiative that even Trump can get behind. But while Eko Atlantic is promising a safer future for Lagos's better-heeled residents, further down the coastline the hazards are growing.

* * *

Seconds after Belgium score their second goal in a 4-1 victory over Russia, Lukman Oladele's beachside bar is plunged into darkness – another breakdown in the fragile electric grid with which all Lagosians are wearily familiar. Evening power cuts used to send Lukman running for his backup generator, propelled by the yells of furious football fans. But this Saturday night, he remains sitting next to me in the gloom, facing the blank television screen, the silence broken only by the bursting of nearby waves. We are the only people here.

The son of a carpenter on the other side of Lagos, Lukman first came to the city's eastern beaches as a boy with his older sister, joining the weekend crowds who flocked to lounge with ice creams and ride scrawny horses on the golden sand. About ten years ago, he decided to start a bar at Alpha Beach, an old fishing village that had become a popular destination for day trippers from downtown Lagos. He found a promising spot and got the word out by distributing handbills to the moto taxi riders. Within a few years, Lukman's place was a fixture of Alpha Beach's humming leisure scene, with ten guest rooms, regular DJ nights and an impressive turnout for his screenings of football matches. But no sooner had Lukman's property expanded than it began shrinking.

Year after year following Lukman's arrival at the beach, the ocean swept in with seasonal surges of unusual power, eating away at the coastline. Between his bar and the beach there had been a stretch of land liberally sprinkled with palm trees, but this was steadily eaten away. The sea swallowed Lukman's guest rooms, and then the rooms where he lived with his family, forcing him to move into a friend's home and send two of his three children to live with their grandmother. When I visit the main bar room remains intact, but its foundations are now exposed by the erosion, and its back wall forms a precarious continuum with the stumpy brown cliff that descends a few feet from floor level to the beach.

At least Lukman's business remains open, if barely so. With the old crowd of visitors now steering clear of the chaos engulfing the village, the rest of Alpha Beach's tourist industry has collapsed. The skyline is dominated by the decaying shell of an unfinished six-storey hotel, abandoned a few months from completion.

I take a walk along the beach – once kept clear of trash by fastidious hoteliers, now strewn with discarded gin bottles and hair extension packets – with Banuso Kehinde Shamusideen, the gangly imam of the area's Muslim community. Lethargic from a lingering bout of malaria, he points out the spot, several metres out to sea, where his mosque used to stand – now lost, along with a Christian church and countless dwellings. Then his arm shifts to the right, pointing a few miles down the coastline where the first towers of Eko Atlantic are silhouetted in sultry haze against the western horizon. There had always been erosion here, the imam says, and it had been getting worse over the past few decades. But he's convinced it was when the Eko Atlantic development began that the situation became critical. Something about that vast intrusion into the sea, he believes, something stemming from the millions of tonnes of sand and rock dredged from the seabed and dumped by the shoreline, has altered the behaviour of the sea at Alpha Beach, accelerating the ocean's advance into the land.

Most of those I speak to over several days at Alpha Beach give the same analysis. While Lagos's whole coastline was already being hurt by the rising and increasingly restless sea, they say, this eye-catching initiative to protect the city's most upmarket district has worsened the damage further down the coast. 'I know about climate change, and that's a big part of our problems,' Lukman says. 'But climate change hasn't done this by itself.'

Ronald is keen to challenge these claims when they come up during a conversation at the sales office on the edge of Eko Atlantic, in a large room dominated by a scale model of the development. The destruction at Alpha Beach was always just a 'matter of time', he says, and locals' perception of accelerated erosion is an understandable response to the fact that it has now started destroying their homes. 'Coastal erosion is not really an issue until it starts affecting where you live,' he says, adding that his group has commissioned research from the Dutch engineering consultancy Royal Haskoning that suggests the concerns about knock-on effects from Eko Atlantic are not justified. Yet an impact assessment produced by the same consultancy, earlier in the project's gestation, predicted that it would indeed shift the coastal erosion eastwards toward places like Alpha Beach.

In search of an outside opinion, I pay a visit to the Nigerian Institute for Oceanography and Marine Research, where I find research director Regina Folorunsho, briskly efficient in a sharp green blazer. Climate change, she says, by raising the sea level and causing stronger waves, has heightened the pre-existing vulnerability of Lagos's crumbly shoreline. The land here consists largely of sediments spat out by the mighty Volta River in Ghana 200 miles to the west, which are then pushed eastward along the coast by Atlantic currents. VI was especially badly hit by erosion, Regina explains, because of a huge barrier that the colonial British built into the sea from the island's western end to protect a shipping channel into the lagoon. That barrier blocked the flow of sediment from the west – while sand kept on flowing from the island to beaches further down the coast. Now, thanks to Eko Atlantic, the erosion of VI has been halted – but this means that places like Alpha Beach are no longer benefiting from the resultant flows of sand. The Great Wall of Lagos, she goes on, seems to have changed wave patterns along the coastline, increasing the force of those sweeping into Alpha Beach.

Her diagnosis raises uncomfortable, complex questions of a sort that will become increasingly familiar in the global struggle to cope with climate change. Were the people around Alpha Beach entitled to expect a continued flow of sand from the west, even if it was created by the progressive destruction of Victoria Island? Was the government right to authorise a sea wall protecting Nigeria's most important commercial real estate, at the risk of creating knock-on consequences elsewhere along the coast? How many other defences against climate change will come with unwelcome side effects – and will the impacts always be skewed in favour of the better off?

There's no question that Eko Atlantic has been a game-changing intervention for VI, Regina says. Indeed, no existing building in the city has benefited more from the new development than her institute, which once sat directly by the island's disintegrating seaboard, and now lies next to Eko Atlantic's western tip. But even in this part of Lagos, the climate risks remain profound, she warns, with increasingly intense rainstorms dumping water into streets surrounded by the rising ocean and lagoon.

Just a few weeks before my arrival, the latest round of severe flooding had hit Lagos, prompted by torrential rainfall of unusual ferocity. For many in the city, journeys home became miserable expeditions, as drivers inched gingerly along roads whose fearsome potholes were hidden deep beneath murky brown water. But in one of Lagos's most ancient settlements the traffic continued to flow, the residents cruising through watery streets much as they had since long before anyone could remember.

* * *

Francis Agoyon's bright cyan caftan sways about him, his house shuddering on its wooden stilts, as he bounds from his stool to mime the struggle he faced last month on a shopping trip to VI, wading through the upscale area's flooded avenues. 'It was flooded to here!' he shouts, his hand slashing partway up his thigh. He grins as he regains his seat, palms planted regally on his knees, before delivering the smug conclusion to the anecdote: 'But here on the water, we don't feel anything.'

Makoko is sometimes referred to as the 'Venice of Africa' – a soubriquet that might seem mocking when one considers the ramshackle look of its wooden homes, the cosmic blackness of its water, spattered with floating islands of refuse so thick that chickens can be seen walking across them. But there is something unique about this place, where thousands of people have made their homes on the lagoon that lies at the heart of Lagos. Just as in its more celebrated Italian counterpart, its watery streets are vital thoroughfares, though instead of stately gondolas they host a congested mass of canoes in constant collision: children paddle by on their way to school, women slowly drift along hawking groceries, while the fishermen cruise in with their catch from deeper waters.

The settlement has graced the lagoon for well over a century – more than two, according to many residents. It was founded by the Egun, migrants from the powerful Kingdom of Dahomey in modern Benin. Fishing people, the new arrivals built their homes on the water to be closer to their catch – and perhaps also to avoid conflict with the area's existing residents. The water dwellers became known for their tradition of *mahoho*, the public shaming delivered on adulterers, who were paraded in boats, their names called out to the beating of a drum – a custom that gave the area its name, corrupted over the years to Makoko.

That long history looked threatened with sudden extinction a few years before my visit, when the narrow waterways filled with boats carrying chainsaw-bearing workers, escorted by armed police. The state government had abruptly decided to demolish a huge stretch of Makoko, deeming it a public hazard. As its waste clogged the flow of water into the lagoon, the state governor said, Makoko's growth was heightening risks of flooding in densely populated areas of the mainland.

Francis's house is filled with the sharp aroma of fish being smoked in an adjoining room by one of the two wives who have borne him eighteen children. Gaps in the floorboards give glimpses of the opaque water a few feet below. He retrieves a small stack of photographs and thrusts them at me theatrically while averting his eyes over his left shoulder. As *alashe*, or supreme chief, of Makoko's Egun, he was in the thick of their attempts to resist the destruction of their homes – an effort that claimed the life of Timothy Huntoyanwha, an energetic member of Francis's cabinet. In the first photograph Timothy lies lifeless in the back seat of a car, a gory mess where a police bullet struck his lower abdomen.

Amid the uproar surrounding Timothy's death, the government

suspended its demolition drive and has showed signs of grudgingly accepting the slum's existence. Recently it allocated official names to Makoko's watery streets, including one named after Francis, where he lives at number 23. Assuming there is no renewed effort to destroy Makoko, Francis says, its people – long used to life on the water – will be far better placed to adapt to rising seas than their wealthier counterparts in places like Victoria Island. 'We are enjoying it here,' he says in his hoarse, insistent voice, 'more than those people in their brick houses.'

Drifting through Makoko's fetid waterways to the sound of crashing canoes, it's easy to find holes in Francis's rose-tinted portrayal of his community. His insistence that there are no serious health problems here, for example, jibes awkwardly with the countless swollen bellies to be seen on Makoko's children, a possible symptom of the malnourishment that can stem from digestive problems triggered by dirty water. His argument that climate change will cause less disruption here, however, is less easily dismissed.

Close to one intersection, I look up to see an elfin figure squatting precariously on a plank, hammering at the second storey of a new building. At 44, Daniel Aide has been working as a carpenter for more than three decades, having followed his father into the trade from childhood. He descends to take a break in a canoe moored to the house, sitting back in a white T-shirt smeared with dust from his work.

He's putting the finishing touches to this home – a two-month job that he's executed with a team of labourers, charging the new owners Janet and Ezekiel about a thousand dollars for the work. Well over a metre above the water, the house's ground floor stands conspicuously higher than most of those around it. That's the case with most of the homes Daniel builds these days. Residents are aware of the rising sea level, and they want their houses – which Daniel says are built to last twenty-five years – to withstand it, Ezekiel says, as Janet fills her canoe with bread, milk powder and instant coffee before heading out for an evening of hawking.

And why not? To equip a Makoko house to deal with higher sea level, Daniel says, adds little to the cost – in almost comic contrast with the titanic engineering that will be needed in places like Victoria Island. All his houses are secured by wooden stilts, driven deep into the silty lagoon floor. To make a house stand higher above the water, he merely needs to use longer stilts.

A few years ago, Makoko improbably exploded onto the global architecture scene, heralded at the Venice Biennale as the theatre for an exciting new experiment in climate-resilient buildings, when the young Nigerian architect Kunlé Adeyemi won the coveted Silver Lion award with his design for a floating school in the community, an elegant A-shaped wooden structure atop a raft of barrels. Adeyemi had spent most of his career in the Netherlands, a world away from Makoko's stilt-lined waterways. Still, his project created welcome work for local carpenters, including Daniel. As the school took shape, funded by international grants, Daniel raised concerns about the materials being used, which differed from those that had been tried and tested over decades by Makoko craftsmen. His objections were brushed aside by those overseeing the project, which was winning enthusiastic international media coverage. Yet after the construction was completed, Daniel's concerns began to look prescient. The new building leaked in the rain and shook in the wind, and was swiftly rejected by the head of the local school, who resumed teaching in his old building. At last, in a morning storm one week after Adeyemi's triumph at Venice, the celebrated new structure crumpled and collapsed upon itself like a shot elephant.

Daniel is interrupted by a sudden splash from a child losing his balance in a tiny canoe, who steadies himself with an oar before paddling guiltily away. If Adeyemi had taken local expertise on board, he resumes, the school might have stood a better chance. Even those designing the latest expensive buildings in Victoria Island, Daniel muses, might get some inspiration from visiting Makoko – not that he rates their chances. As the sea rises, he predicts, Lagos's developers and urban planners will struggle to deal with the water intruding into the low-lying areas that are the beating economic heart of the city, and indeed of all Nigeria. 'They don't understand these things,' he says, his face illuminated by glimmers of late-afternoon sun bouncing off the rippling lagoon. 'But us – we live with the water.'

CHAPTER 5

Climate change has emerged as the greatest threat to
coral reefs . . . Even achieving emissions reduction targets
consistent with the ambitious goal of 1.5°C of global
warming under the Paris Agreement will result in the further
loss of 70–90% of reef-building corals compared to today,
with 99% of corals being lost under warming of 2°C
or more above the pre-industrial period.

– Intergovernmental Panel on Climate Change,
Global Warming of 1.5°C. An IPCC Special Report (2018)

BAA ATOLL, MALDIVES

Our flippers propel us through an aquatic Pompeii, a once thriving, kaleidoscopic jungle of coral now frozen in grimy suspended animation. We pass the monochrome remains of sprawling table corals, stretched out from a narrow base like giant mushrooms, now gradually teetering toward the seabed. There are grey formations reaching up in clusters of muscular fingers, others like bundles of twigs gathered from a forest floor, more sections that have degraded into amorphous lumps of rubble.

Once every couple of minutes Sendi spots a tiny sign of life – a fist-sized colony of coral, obstinately alive – and he shoots down to photograph it. But those moments come as rare respite from the blanket of lifeless limestone, caked in algae and bathed in the sand formed by its own disintegration.

Back in our boat, rocking gently in the 25-mile-wide lagoon of Baa Atoll, I ask Sendi, the Maldives' most experienced and best-known diver, whether the reef shows any sign of recovery. A slight man with any hint of excess fat lost to his forty-seven years in the water, known to no one by his legal name Hussain Rasheed, Sendi is usually a sprite-like source of mischief, sea-blue eyes gleaming over glasses perched on the end of his nose, crunching out conspiratorial sniggers between puffs on a Camel Light. But now he looks serious.

'It's like a graveyard,' he says, seawater still dripping down his cheeks into a dense white beard. 'Like a horror film.'

Sendi swam through these waters when they were still carpeted with coral polyps, minuscule animals shaped like sea anemones that cluster together in colonies. Beneath themselves the polyps formed the rock-hard reef, a shared exoskeleton of limestone that broadened and thickened with every generation. Each polyp had its own tiny hollow on the reef and sheltered there during daylight, before extending minute tentacles at night to snare passing plankton.

The reef's riot of colour came not from the transparent polyps but from their life partners, the *zooxanthellae* – single-celled beings that lived inside the polyps' tissues. The corals got most of their nutrition from sugars generated by their unicellular lodgers, which in turn received shelter and the carbon dioxide needed for photosynthesis. It was a relationship of stunning beauty and simplicity, forged over hundreds of millions of years of evolution. But on this and countless other reefs across the Maldives, it came to a brutal end over a few weeks in 2016, a year when the global average temperature hit yet another modern record.

The underwater tragedy unfolded before Sendi's eyes. 'It's a strange feeling for a diver, to swim in a lagoon that's as warm as bath water,' he says. 'And then the coral just turned white.'

As the Maldivian summer began that March, sea temperatures surged well above 30C, far beyond the corals' narrow comfort zone. The polyps went into a state of stress, and the convulsions of their tiny tissues expelled the zooxanthellae into the ocean. For some days, the corals clung to life, using their tentacles in a desperate nocturnal hunt for passing scraps of nutrients. The now colourless polyps had all but vanished before they died, leaving as a ghostly legacy tonnes of colourless, intricately fashioned limestone.

* * *

Nearly two centuries before Sendi and I sat bobbing in the lagoon, with $1,000-a-night resorts in all directions, a 42-year-old English sailor named Robert Moresby passed through it in the ageing sloop-of-war *Benares*. He was on a mission of high national importance: to map the Maldivian reefs.

The motivation was purely economic – the jagged reefs were a grievous risk to ships plying the burgeoning Indian trade routes. But the work made a major contribution to the science of coral systems, providing crucial data for Charles Darwin's groundbreaking first book, *The Structure and Distribution of Coral Reefs*. The Maldives' round atolls were the result of volcanic mountains gradually sinking into the sea floor, Darwin wrote in the 1842 tome. The corals surrounding the mountains had grown upwards, keeping a constant depth from the sea's surface, their limestone base getting thicker every year – even after the mountains in the middle had long vanished under the ocean.

Darwin's work provided a basis for all modern research on this subject. Seismic mapping of the Maldivian atolls has shown that they rest on limestone foundations more than a mile deep, built by successive generations of corals over fifty million years. Contemporary scientists have confirmed, too, Darwin's assessment of the immense amount of sand that passes from the digestive tracts of parrotfish, whose sharp teeth bite off pieces of limestone as they scour reefs for algae. Most of the stunning white sand in those slick Maldivian resort photographs, it turns out, is fish shit.

Yet there was one crucial angle that Darwin missed: climate change. Much more than volcanic subsidence, we now know, the Maldivian atolls were shaped by the huge sea level rise that occurred as the last ice age drew to a close. Over fifteen millennia from 20,000 BC, the oceans rose by nearly 120 metres – and the corals rose with them. As the temperature stabilised, so did the sea level – and as the centuries passed, the 1,100 islands we now call the Maldives were formed from fragments of coral skeleton collecting in huge mounds within the circular reef atolls, an idyllic home for human settlers who first arrived about 3,500 years ago.

But even at the time of Moresby's survey, it seems Maldivians were painfully aware of their precarious geography. 'A singular tradition exists among them,' a sailor named Boyce wrote in his chronicle of the mission, 'that all their islands are to sink down into the sea, and that large ships will sail over the places they now occupy.'

Whatever the origin of that tradition, it uncannily anticipated the threat that now looms over the Maldives' long-term future. Today, man-made warming has begun to drive sea level rise that will be even faster than that which ended the last ice age. And unlike during that prehistoric episode, reefs almost devoid of living coral will be unable to grow in line with the rising oceans, leaving the sandy islands within the atolls increasingly exposed.

The 2016 disaster in Baa Atoll was part of the worst wave of coral bleaching in recorded history, which swept the tropical oceans from 2014 to 2017, devastating colonies from Florida to the South China Sea and killing half of Australia's 1,400-mile Great Barrier Reef. Covering two thirds of the earth's surface, the ocean has absorbed most of the heat created by carbon emissions, which has increased its average surface temperature by about 1°C over the past century. With coral's extreme sensitivity to small temperature rises, bleaching – a once rare phenomenon – is occurring with increasing violence and frequency. Reefs are now providing a spectacular first look at the devastation that climate change is set to wreak on global ecosystems.

In a landmark report a few months before my arrival in the Maldives, the Intergovernmental Panel on Climate Change warned that global warming was on course to kill off virtually all the world's reefs by 2100. 'The predictions of back-to-back bleaching have become the reality,' the IPCC said. Warming of 2°C – a level that will be easily surpassed this century on current trends – would cause the death of 99 per cent of the world's corals, it went on. Even with a rise of 1.5° – a target requiring hefty emissions cuts, now being chased by more ambitious governments – up to 90 per cent of the world's coral would die, the report warned, a disaster that would 'remove resources and increase poverty levels across the world's tropical coastlines'.

All ecosystems will be affected by climate change, but these dazzling underwater jungles are set to vanish at a brutal speed, virtually an instant on the epic timescales of coral reefs. It means the loss of some of our planet's most astounding beauty, and a withering economic impact. Reefs support a quarter of the world's marine species and millions of human livelihoods – notably in the Caribbean, and the 2 million square mile 'Coral Triangle' that runs from Malaysia to the Solomon Islands.

No place on earth has more to lose from this catastrophe than the Maldives, a country literally built on coral. The reefs have offered vital protection for the islands against tidal waves and storm surges, reducing the threat of flooding and erosion. Maldivians credit them with saving the country from total devastation when a massive tsunami struck the region in December 2004. But now the reefs that created and guarded the Maldives are dying, beneath a steadily rising ocean swept by increasingly powerful storms. For this nation of 400,000 islanders, the world's lowest-lying state, it could prove a fatal cocktail.

* * *

On a narrow street in central Malé, the Maldivian capital, I edge my way through an angrily stuttering flow of scooters and pickup trucks towards a modest commercial building that looks to have been under siege. Flanked by a DHL office and a baby products shop, its windows have been smashed and hastily patched, and the glass in its front door replaced with plywood. From its yellow walls protrude two loudspeakers, and between them hangs a banner covered with photographs of politicians, centred on the face of the man I have come to see: Mohamed Nasheed.

As the Maldives' president, Nasheed became the first political superstar of the global climate fight with his emotional appeals for action at the 2009 Copenhagen Summit. His most celebrated move came a few weeks before the climate conference when he and his cabinet donned scuba gear to hold an underwater crisis meeting. The stunt highlighted the danger that the rising ocean would consume the Maldives, where 80 per cent of the country is less than 1 metre above sea level – the height of a 4-year-old child.

Just two years after Copenhagen, Nasheed was ousted in what he called a military coup, and was imprisoned before fleeing into exile. But he got a second chance. I meet him two months after he walked with cheering supporters through the streets of Malé, returning home after a shock election result. Banned from running for the presidency himself, Nasheed had thrown his weight behind his childhood friend Ibrahim Solih, who swept to power. As the leader of Solih's Maldivian Democratic Party, Nasheed is now back in a position of high influence – and he is resuming his battle to save his country from climate change.

Upstairs from the battered entrance – testament to the fiery tensions of the election campaign – Nasheed is at a computer in the MDP's conference room, neatly turned out in silver cufflinks and a tie to match the party's yellow banners. It's a few weeks after he made a high-profile return to international climate negotiations, leading the Maldives' delegation to the UN global conference in Katowice, Poland, where he grabbed media attention by proclaiming that his country was 'not prepared to die'.

But a decade after he hit the global climate stage, Nasheed is horrified by how little has changed. For all the promises from world leaders, carbon emissions have continued to rise, keeping us firmly on course for dangerous levels of warming. Meanwhile, scientific research on the extent of that danger has been forging ahead – nowhere more so than on the world's ailing coral reefs.

Even as he keeps trying to fend off the rising sea, pleading with the world to cut its carbon emissions, Nasheed knows he also needs to step up the Maldives' readiness for climate change. And the most important way to do that, he has now concluded, is to save the coral.

'Sea level rise is one thing,' he tells me, his gaze intense behind his rimless spectacles. 'But sea level rise without the reef is a disaster.'

Before heading to Poland, Nasheed made a visit to Richard Branson's private Caribbean island, where the British billionaire promised to support his plans for a new organisation to protect the Maldivian coral. Nasheed leaps from his seat to show me a presentation on artificial reefs, with coral growing on pyramidical 3D-printed frames, which he hopes to install across the Maldives to protect islands from surging waves. Such initiatives risk being blown up on the runway, however, if the fledgling corals are devastated by bleaching events in the warming waters. So the time has come, Nasheed believes, to 'look at how we may be able to genetically interfere'.

Coral spawning is surely the animal kingdom's most spectacular sex. Once a year, soon after a full moon, a reef explodes in a synchronised flurry of eggs and sperm, propelled in tiny pearl-shaped bundles from the polyps' flexing mouths, rising in a multicoloured upside-down blizzard to the sea surface. There, over the course of a week, they fertilise in their millions, the larvae at last sinking to the sea floor in search of hard surfaces to latch onto. But for all its splendour, the natural mating process looks insufficient to ensure the species' survival, and the recent disasters have added new energy to a wave of scientific intervention.

After the latest round of bleaching, researchers at the Australian Institute of Marine Science went diving on the Great Barrier Reef in search of living coral, which they detached with hammers and chisels and took back to their labs. By interbreeding these unusually hardy survivors, they're hoping to create new hybrid strains that will be better able to cope with the twenty-first century's rising sea temperature. In Hawaii, scientists have been subjecting corals to near-death experiences in tanks, exposing them to temperatures just short of lethally warm, to see if they can be gradually conditioned to survive such ordeals. At California's Stanford University, researchers have started using the newly developed CRISPR technology to edit individual sections of the coral genome.

Nasheed's new foundation will step up efforts to put this emerging science into practice, and build ties between scientists and governments of countries that have the most to lose from the devastation of the reefs. It might seem hard to believe that this work could do much to temper the epic destruction forecast by one study after another, given the enormous size of the threatened reefs and the extensive warming already locked into the global climate system. But Nasheed is still brimming with the zeal of the freedom fighter he once was. In his youth, he spent eighteen months in solitary confinement for activism against the country's long-running dictatorship. Guards on his prison island put crushed glass in his food and left him chained to a chair in the blazing sun for days on end. Having survived to become his country's first freely elected leader, he now brings his old defiant fervour to the battle to save an entire global ecosystem, which may hold the key to his country's survival.

'I have a view that you don't have to succumb, you don't have to give in,' Nasheed tells me. 'None of these reefs have to die. We still have a window of opportunity.'

* * *

Before leaving the Maldives, I spend a week without shoes on Fulhadhoo, a tiny island in Goidhoo Atoll, a half-hour speedboat trip to the south of Baa. A thin spit of land one mile long, it has no paved roads and a population of about 180.

Some were sent here as convicts, under an ancient Maldivian tradition of banishing criminals to distant islands. These include Joachim Bloem, a young German tourist sent here in 1978 after killing his girlfriend in Malé. Bloem has built a new life as an islander, converting to Islam and taking the name Ismail, chatting with his growing brood of grandchildren in accentless Dhivehi. Now he's making his small contribution to the restoration of the Maldivian corals, creating hundreds of marine concrete frames used to build artificial reefs, sites for breeding hardier strains of coral that survived the bleachings. 'It can help, for sure,' he tells me quietly one morning, over canned coffee in his outdoor workshop.

Fulhadhoo's northern edge is rimmed with ancient sediment hardened into a jagged rocklike surface, over which black-shelled crabs scuttle away their mornings. On the opposing side of the island, a mud path leads through long-leafed undergrowth overlooked by thin, twisted palm trees, before suddenly bursting onto a ludicrously perfect beach of feather-soft white sand that turns raspberry pink at dusk. The island has so far been spared the attentions of the millionaire resort developers who dominate the Maldivian economy. And its surrounding reefs have come through the recent bleaching events in far better shape than most others in the Maldives, thanks to faster currents that dispersed the warmer water before all the corals were killed.

This is where Sendi has come to spend his retirement, watching over the dive school run by his daughter and son-in-law. One day we strap tanks to our backs and swim down to Fulhadhoo's finest reef, a short distance from its southern shore.

The scars of 2016 are visible, with some dead sections covered in sand. But resurgent life is everywhere. Here are slabs of blood-red coral, creased and undulating like giant slices of brain; colonies shaped like indoor cacti and others like giant pin cushions. We drift past studded protrusions like tentacles with glowing blue tips, and giant centuries-old boulder shapes with ridges forming orange mazes of endless complexity. And at every turn is a technicolour crowd of fish – the surgeons, with their royal blue flanks and long yellow mohawks, the gothic black damsels with their golden tails, the parrotfish grinding sand in their pink-striped bellies.

The reef may have survived the last round of bleaching, but it's becoming ever more endangered. March is just a few weeks away

– the month that has become imprinted on the consciousness of Maldivian marine biologists, when water temperatures start to surge and the now annual threat of mass coral death rears its head.

Yet Sendi is upbeat as we head back to shore, reinvigorated by our brush with the resplendent corals. He grew up in a house made of coral: one of his earliest memories is of watching his father build it, using blocks of reef limestone, and cement made from ground-up branching coral mixed with coconut syrup. He's relied on the reefs for his livelihood, no less than the fishermen who have harvested their surrounding waters since the Maldive islands were first inhabited. He's in no mood to accept the possibility of their extinction – let alone that of his entire nation.

'Nature is fighting back,' Sendi assures me, grinning once more, as our boat eases through the shallow emerald water towards Fulhadhoo's wooden jetty. 'The coral will save us.'

CHAPTER 6

In general, societies most at risk from climate change – and thus most in need of active adaptation – are those that are least responsible for emissions.

– David G. Victor et al., *Mitigation of Climate Change. Contribution of Working Group III to the Fifth Assessment Report of the Intergovernmental Panel on Climate Change* (2014)

NUATAMBU, SOLOMON ISLANDS

Wading through a warm, shallow stretch of the South Pacific, soft coral sand underfoot, Chief Maso Sambolo points out submerged sections of what used to be the heart of his island. To our right, a volleyball court. Ahead, a community hall. And, scattered right across this narrow channel, passed over by children in dugout canoes, the wood-framed homes of about twenty families, vanished but for a few pieces of timber still lodged in the sand.

The Solomon Islands were named after the prodigiously wealthy biblical king, by sixteenth-century Spanish sailors who reached them from Peru, driven by Inca tales of gold from the western oceans. Today the name carries an ironic tinge: the people of these 900 islands, stretching between Vanuatu and Papua New Guinea, have among the lowest incomes in the Pacific region (with little mining to speak of). And now they face a new challenge – sea level rise, at a pace greater than almost anywhere else on earth.

In the past few decades, the ocean has been rising around this archipelago by nearly a centimetre each year. That's about triple the global average rate, with scientists attributing the faster pace here to changes in wind patterns, which in turn have been linked to global warming. As the University of Queensland's Simon Albert – one of the few academics who has studied the subject deeply – told me shortly before my visit, the Solomon Islands offer a valuable window into the effects of sea level rise at a scale the rest of the world will experience before long.

For Maso's people, the impact has already been profound. Maso is a large man who speaks in a mellifluous baritone, and when he laughs it's not a private expression of amusement but a whooping celebration to which everyone in earshot is invited. He's followed a long line of ancestors to head this island settlement, just off the northern coast of Choiseul, the westernmost of the Solomons' six major islands. He took charge of Nuatambu at the age of 33, in 2014 – the year the water finally closed over its sandy central section. Maso found himself the leader of two islands, the first chief of a transformed, divided Nuatambu, with many of his people now dispersed along the thickly forested northern edge of the mainland.

The Solomons have been inhabited for at least three millennia, but have been a single nation for a tiny fraction of that time. Over the preceding centuries, fighting between tribes produced a fragmented scattering of extremely isolated cultures, their settlements hidden in the hills for protection against invasion and head-hunters. One 1943 journal article catalogued six different languages on Choiseul, which then had a population of about four thousand. By then, however, inter-tribal violence was over, suppressed by the twin influences of Christian missionaries and ruthless British forces. Across the archipelago, people descended from the mountainsides to the more accommodating coastal areas. Tiny warlike communities, concealed high in the forest, became fishing villages with burgeoning populations, trading peacefully within and between islands. But in places like Nuatambu, people are now being forced back to the hills.

Peter Navala is the oldest man in Nuatambu, and one of the last residents of the low-lying sandy section on what is now the western island. At 77 his blue eyes remain bright and lively above sharp cheekbones that cut back towards the thick white curls at his temples. Peter is living with his son's family in one of the few remaining houses still standing on this flat beachfront section of the island, built on wooden stilts with thatched walls of dried leaves, a canoe lying in the space beneath it. But beyond the house's corrugated metal roof, I can see the new home Peter's son is building, clinging to the hillside amid soaring forest that, from this distance, forms an unbroken blanket of tropical green. It will be difficult up on the hill for an old man when the rains come and the paths turn to mud, Peter reflects. But he doesn't question the need to move. The canoes cruising over the old village centre serve as a daily reminder of the sea's threat.

Convincing the last residents of the sands to move to the hills has been Maso's biggest mission as Nuatambu's chief. The erosion is continuing unabated, with the shoreline of this amputated western island moving steadily towards Peter's home and the handful of others still standing near it. The king tides of December and January rise higher each year, sometimes flooding the whole area. It's no longer safe there, Maso says. Better that the last residents move now, to avoid becoming suddenly homeless when the ground beneath their houses gives way.

My journey to Nuatambu began with a dawn flight from Honiara, the national capital, on a small propeller plane that took me to Taro, the provincial capital of Choiseul. There I chartered a small boat that carried me eastward for four hours through bouncing waves and cold, piercing rain. To my right extended the northern coastline of Choiseul, draped with closely crowded trees that stretched towards high volcanic ridges, interrupted only occasionally by small clusters of houses.

Maso happened to be standing by the eastern island's wharf when I arrived, inspecting work on the frame of a new village hall to be built on its hill. He had no objection to my staying for a few days and studying the situation, but this was a matter that would need formal approval from the community. He blew a large conch shell to summon a meeting, and I stood on the wharf explaining my plans to a semi-circle of about forty villagers, men to my left, women to my right. The initial response was dubious. Nuatambu's people viewed the erosion of their island as an outcome of environmental harms committed by countries far larger and wealthier than theirs, which had done nothing to assist them. Some of them saw no prospect that talking to me would help matters. When they voted at the end of the meeting, however, the verdict – albeit not unanimous – was that I could stay and hear the stories of this beautiful, damaged place.

The wary welcome chimed with the name of the settlement, which translates as Forbidden Island (tambu, or taboo, is a word found in various South Pacific languages, and was incorporated into English after British sailors reached here in the 1770s). During my stay in Nuatambu I hear compelling tales of the place's history, most of them impossible to substantiate. When people still lived up in the hills, I am told, this was a site for human sacrifice, strewn with severed heads. Later, the island became the centre of production for *kesa* – hollow discs carved from thick clam shells that served as currency across Choiseul, and are still used to seal marriage agreements and settle disputes. The knowledge of how to make *kesa* was already long forgotten in 1943, according to that old journal paper. Islanders tell me the currency was made by a foreigner called Pong – dubbed Pongo by locals – who had stayed in Nuatambu for a few years, centuries before. A stranded Chinese craftsman? Pongo got rich from his production of *kesa*, and died leaving a stash of gold somewhere on the island that has never been discovered. A couple of locals confide privately that the nervousness about curious outsiders stems mainly from the risk of losing this buried treasure.

For all the richness of Nuatambu's oral history, nothing in it is more dramatic than the events of the past decade. The central section had been shrinking slowly for years, the sea steadily rising up its delicate slopes, but the erosion picked up pace around 2011. One by one, the buildings at Nuatambu's heart gave way as the water spread beneath them and washed away their sandy foundation, their supporting stilts leaning and straining towards final collapse.

Weary Barivodu's home was one of the last to fall in 2014. She'd seen the disaster coming and had already built a new house up on the hillside. When her home on the sands started to give way, Weary began moving her belongings to the new dwelling, but not quickly enough. The old house's final collapse came suddenly, with a large wave that swept away a valuable axe, some treasured sewing patterns and a trove of *kesa*.

What happened next was no less distressing. Weary found herself trapped in a dispute with her new neighbour, a cousin who insisted that she had no right to the land where she had built her refuge. Unable to tolerate the hostility, Weary abandoned the plot and moved across to the mainland, where she now lives in the home of her sister, who is away working as a teacher in another village. In a loose blue sleeveless blouse, she sits on the house's raised veranda, the distant sound of children's voices permeating the palms dangling orange-skinned coconuts. Before long her sister will retire, and return to live in this small home. Weary has no idea where she will go then.

Weary's land dispute seems to reflect a broader straining of Nuatambu's community bonds by its sudden physical division. On Saturday – the sabbath for the Seventh Day Adventist faith to which the whole of Nuatambu belongs – the small beach on the eastern section becomes a parking lot, crowded with slim canoes of varying sizes, each carved from a single tree trunk. Up a steep muddy staircase formed by the roots of ancient trees, the church tops Nuatambu's eastern hill, commanding a view over miles of coastal waters underlaid by sea grasses and coral gardens.

'It is good to see you all here,' says Gandly Galoghasa, one of the community's leaders, beginning the morning's sermon to the crowded church, his usual beachwear swapped for a shirt and tie. 'The sabbath is the only time we all get together, now that climate change has scattered us.' The church activities continue throughout the day, with hymns sung in well-practised harmonies, concluding at sunset with a fervent call by one senior figure for volunteers to assist in the renovation of the school. Such community work has always been a central part of village life in Nuatambu, but the turnout has been slipping since the island's division. People just spend less time together now that they need canoes to see each other, Maso says. The attendance at his meetings has dropped, too, with many of those now living on Choiseul unable to hear the conch shell's call.

Gandly is among those who have been forced to the mainland, his old home long since lost to the waves. One day we walk to the site where it once stood, the ocean lapping above our knees. The water is littered with the bleached remains of large fallen trees; children are splashing and playing amid them, their shrieks carrying far across the bay. The kids are spending much more time in the water now that there is so much less open space on land. For them, all this must seem normal, Gandly says quietly. They will never quite understand what has been lost.

* * *

The engine silenced, our small craft drifts towards the bank of the slender jungle river. Around us, dangling lofty arcs of creepers, rises the mangrove forest, as tall as a built-up inner-city district and much more densely packed, with little light penetrating to its lower reaches. A pair of large hornbills fly over the river as from somewhere far to my left comes the whine of a lone chainsaw, part of a Malaysian-owned logging operation. And to my right, on 500 hectares stuffed with more vegetation than air, is the site earmarked for the new capital of Choiseul province.

My companion on the boat, a scholarly type cradling a handheld location tracker, is Isaac Lekelalu, a civil service veteran leading the extraordinary initiative to move an entire provincial capital to a site more resilient to climate change. 'This is a choice we have to make, in order to survive,' Isaac says. It's one of the first projects of its kind – and may hold important lessons for others to follow in the decades to come, as coastal settlements on every continent come under growing threat from rising sea level.

It's an accident of history, Choiseul provincial officials keep stressing to me, that the small coral island of Taro, 65 miles north-east of Nuatambu, became their capital in the first place. Taro's flatness – in contrast to the hilly major island of Choiseul – made it a good site for a US military airstrip during the bloody Solomon Islands campaign of the Second World War. (A 1944 *New Yorker* story chronicled the epic survival story, a few hours to the south, of a shipwrecked young naval lieutenant named John F. Kennedy, who scrawled an SOS message to base on a coconut that would later sit on his Oval Office desk.)

The grass airstrip, and the modest infrastructure that developed around it after the war, meant Taro became the administrative hub when Choiseul was granted provincial status in a 1992 shake-up. But the province's new government was keenly aware of Taro's tiny area – just half a square mile – and immediately started discussing an expansion to the nearby mainland. For years, these plans languished, with national leaders showing little interest. But now, the Choiseul leadership is pushing with new urgency, not just for an expansion but for a wholesale relocation of the township, home to about a thousand of the province's 26,000 people. The driver is the realisation that thanks to sea level rise their island, already too cramped, is shrinking.

The relocation has become an obsession for Geoffrey Pakipota, Choiseul's provincial secretary, a grave, white-haired 49-year-old who started work on the plans at 23. I find him in the yard behind his home on Taro's eastern edge, taking a rest day after a week of meetings in the national capital of Honiara to seek support for the project. Geoffrey rises from his hammock to show me the shoreline erosion closing in on his and other houses. Just a few paces away, the cracked grey earth gives way to a mass of roots, tangled like electrical wires, that protrude over a tiny beach of stones lapped by the water. Several metres out to sea is the site of the old marketplace, one of a growing number of buildings lost to the waves.

Over the past few years, Geoffrey has spearheaded the new

momentum brought to the relocation drive by Choiseul's government. In 2016 it completed the $1.2 million purchase of territory on the mainland. Shortly before my visit, it hired surveyors to carry out an advanced zoning study for the new town, and engineers to start preparing its main road. Every one of the Taro residents I speak to supports the relocation plan. They no longer feel safe in this island, too small and getting smaller.

The full cost of building the new capital, however, is far beyond the modest provincial budget. Even the national government can hardly afford to fund Geoffrey's plans, which will cost – if a planned new port is factored in – hundreds of millions of dollars. The gross domestic product of the whole country is just $1.4 billion, a bit over $2,000 per inhabitant. But if the national leadership throws itself behind the relocation project, the Choiseul government hopes, it could win some of the billions in funding that major economies have promised to help developing nations adapt to climate change.

So, the week before my arrival, Geoffrey and Isaac and most of their senior colleagues pursued an exhausting series of meetings in Honiara, lobbying Prime Minister Mannaseh Sogavare and all his main cabinet ministers for assistance. A breakthrough remained elusive, and the frustration of the endless campaigning is now showing. 'The national government has been too slow,' Geoffrey says. 'They are not innovative enough.' In Nuatambu I saw how the effects of sea level rise could pull a small community apart. Geoffrey now warns it could have a similar effect at national level, in a country that is highly fragmented both geographically and culturally. Papua New Guinea, ten times the Solomons' size in population and economic clout, is less than thirty miles to the west, the mountains of Bougainville easily visible across the sea. If Honiara keeps neglecting the needs of this distant province, Geoffrey says, 'there is the question of whether Choiseul will still be part of the Solomon Islands.'

In theory, the solution to Geoffrey's conundrum could come from the Green Climate Fund. This global body was set up to deliver on one of the vital principles of international climate talks: that rich countries, which bear the overwhelming share of responsibility for carbon emissions to date, should help poorer ones to deal with the fallout. On the day I arrived in the Solomon Islands, the GCF and the World Bank signed a financing agreement on a major new hydroelectric plant to power Honiara, shifting the capital city away from its reliance on diesel generators. With $86 million in GCF funding, it's one of the single largest projects to get backing from the fund.

Officials in Honiara and Taro say the Choiseul relocation seems like precisely the kind of thing the GCF ought to be supporting – but they don't show much hope that it will do so. The rich nations are still falling short of their promise to provide $100 billion per year in climate assistance to poorer countries. And beyond the headline numbers, there's also the question of where the money is being directed. Of the cash so far allocated by the GCF, much more has gone to fund renewable energy projects in developing countries, like the hydroelectric station in the Solomons, than to assist them in coping with the impacts of climate change. A power plant proposal is far easier for a financial body to evaluate than one to move a thousand people. But as sea level rise and other climate impacts worsen in the years ahead, the question of how to help those displaced will become increasingly urgent.

In small communities like Choiseul, international bodies have an opportunity to gain experience in supporting the kind of relocation work that will over time be needed in coastal areas throughout the world. Until that funding comes, the Taro township move will remain the pipe dream of an increasingly desperate provincial government. 'It doesn't matter where the money comes from,' Watson Qoloni, Choiseul's premier, tells me in the administration's tiny office complex, maps of the future township pinned to the wall behind him. 'Whether it comes from God or Satan, we'll take it.'

* * *

The dolphins drew the sea people to Fanalei, an uninhabited strip of coral sand and mangrove trees perched in easy reach of South Malaita, one of the larger islands on the eastern side of the Solomons archipelago. They came in dugout canoes from their homeland far to the north, where the language was Lau, a tongue unintelligible to the Sa'a speakers on the forested mainland near their new home.

Those bush people, as the Fanalei islanders call them today, had been slow to exploit the stunning abundance of dolphins in their waters. That was to ignore a gold mine in Malaita, where dolphin teeth served the same monetary function as Pongo's *kesa* in the west. So the northern newcomers set up their outpost on Fanalei and began slaughtering dolphins in their thousands: forming semi-circles of a few dozen canoes from which they banged rocks together under the water, driving the panicked beasts into the muddy shallows, where they were wrestled into submission and slaughtered.

Over the ensuing centuries, the Fanalei islanders maintained a delicate coexistence with the bush people, at a time when the Solomon Islands' scattered tribes were still riven by violence. Sa'a villages hired Fanalei's potent fighters to perform contract killings of rival leaders, granting pieces of land on presentation of the heads. These missions shored up Fanalei's food supply, as they amassed a hilly plot on the mainland on which to grow crops that would supplement their fish diet.

The Fanalei islanders never thought of building homes on that farmland, unpleasantly muddy in the rainy season and plagued by malaria-carrying mosquitoes. But now they have been forced into a dramatic reassessment. The swelling Pacific has destroyed the densely populated southern end of Fanalei, including its church, of which only the concrete foundation still remains. Most of those who lost their homes have now moved to Fouele, a new settlement built on the islanders' mainland plot, dividing the community for the first time since their arrival from North Malaita centuries before. For a few months at the start of each year, the men of Fanalei and Fouele still come together for what may be the single biggest dolphin hunt in the world, with up to 1,600 kills a year according to one recent estimate. But while the dolphin hunt is as fruitful as ever, the community pursuing it is in crisis.

Aside from a couple of older men with a going-down-with-the-ship mentality, all the remaining inhabitants of Fanalei seem to be eyeing an escape to the mainland. Among them are Kingsley and Suzanne Ouou, a couple in their mid-twenties with two small children, who speak to me outside their home on the island's western side, its walls and sloping roof of thatched leaves harvested from the region's luxuriant sago palms. Kingsley views the prospect of life on the muddy mainland with listless resignation. Suzanne is far keener. Island life can be tough for the women of Fanalei, who typically come from other parts of Malaita, and are expected to fetch water from the mainland during the long dry season, paddling back and forth bearing plastic jerrycans. 'I had no idea about all these problems before I married Kingsley,' says Suzanne, a quietly humorous character with her hair tied in a bun. 'Maybe in the future, this place will be completely washed away and the only option will be to move.' But for those still living on Fanalei, that escape route cannot be taken for granted. Under the stress of sudden migration, relations with the bush people are fraying.

Down a muddy path on the fringe of Fouele, perched by a swamp and almost hidden among the mangrove trees, Elias Pilua has built his new house, the old one having been destroyed by the same king tide that wrecked Fanalei's church. This mainland home is built in the same style as those on Fanalei, raised on posts fashioned from tree trunks, but it's much further from the waves, and Elias is no longer troubled by fears of erosion. Instead he has a new concern.

Elias guides me up a slippery path, hacking at undergrowth with a long blade, into the hills where the villagers grow their crops, wide green leaves bursting from mounds of red earth on slopes that seem absurdly steep for agriculture. The crops are mainly cassava, yam and taro, grown almost entirely for a family's own consumption. Small amounts of sugar cane and banana are also grown, and can be sold for money to buy things like soap and cooking oil. This cash income is meagre – Elias's monthly average of about 30 US dollars is typical.

Even this modest lifestyle is under threat, says Elias, a wiry man in his mid-forties with precise English honed by two years of theology studies in North Malaita. As Fouele has grown, it's begun to face increasing hostility from some in the neighbouring communities. 'They tell us that we are not from this place, we are not owners of this land, and we should not be working it,' he says.

The main antagonist for Fouele's people is a man named Robert Tehena who lives in Kalona, a nearby Sa'a village. He claims that their new settlement has expanded far beyond whatever territory they rightfully acquired, onto land that had been owned by his family for countless generations. Most recently he's issued a complaint seeking to block the building of a new community hall, which is proceeding nonetheless. The centre of the village is filled with the sound of its construction, with a dozen Fouele men sanding planks, installing window frames, piling mounds of cement. While the villagers have decided to forge ahead with the hall – it's on land they already occupy – they're nervous about testing Robert's patience by expanding their territory, which they say is now too crowded to accommodate more arrivals from Fanalei. The chronic dispute is casting a shadow over any plans for the village's future.

The journey to Robert's village takes me northbound in a motorboat along the fringes of the sprawling mangroves, then deep inside them, paddling now through water a few inches deep, between the black roots that probe above the surface like alien tentacles. Next a half-hour on foot through the forest, the silence unbroken but for the squeaks of crickets and the belching of the mud that oozes between my toes and at one point swallows the lower part of my right leg.

When I find Robert in the centre of Kalona, a scattering of houses on a grassy hilltop carved out from the forest, it's as though he were deliberately playing up to the villain's role in which he's been cast by the Fouele villagers – hunched and scowling, a heavy machete in his right hand. But he's eager to explain his position, vanishing to reappear ten minutes later with a yellow binder full of legal documents.

Even before half the Fanalei people came to live on the mainland, he says, the community had been trespassing on his land. His documents seem to indicate the law is on his side. He's taken the case to a panel of chiefs, where his claim to the land was upheld. He's since obtained an eviction order, but has not yet enforced it. He feels sorry for the islanders, he says. He knows they have nowhere else to go. But why should he be the one forced to pay for their resettlement, he asks, by giving away his most valuable asset?

If climate change is the cause of the Fanalei exodus, Robert's share of the blame is smaller than most, his carbon footprint negligible compared with my own or any Westerner's. Like most Malaitans, he's not even connected to an electric grid. He's willing to sell the land for a fair price, he says. And since the villagers' subsistence economy leaves them with virtually no cash to buy it, he adds, the most just solution would be for the world's rich countries – the biggest drivers of the problem – to supply the funds.

'Those countries are responsible, but they are not taking responsibility,' Robert says. In Fouele, too, villagers propose this as a logical answer to their predicament, perhaps their best hope of escaping the legal limbo undermining their attempt to rebuild their community. It's a rare point of agreement between the two sides – and one that I feel in no position to dispute.

CHAPTER 7

Sea surface temperature changes and sea level rise, both caused by temperature changes, will directly affect Bangladesh, perhaps more than any other non-island nation . . . The number of climate migrants is projected to increase in all scenarios by 2050 . . .

– World Bank, *Groundswell: Preparing for Internal Climate Migration* (2018)

GABURA, BANGLADESH

The tiger widows break into peals of pitch-dark laughter when I ask what future they see for Gabura, their tiny island nestled between two rivers in the far southwest of Bangladesh. 'There won't be a Gabura in thirty years,' says Shahida Bibi, the most talkative of the five, with lively eyes and locks of white hair peeking out from under a jazzy red headscarf. 'This place is headed for ruin.'

The schadenfreude is perhaps understandable after the treatment she and the other women have received. Their husbands were killed by some of the hundreds of tigers that prowl the nearby Sundarbans, the world's biggest mangrove forest, which stretches across great swathes of this region into the Indian state of West Bengal. An ancient local tradition holds women widowed by tigers to be cursed, even somehow culpable for their husbands' demise. Shahida and the other women became outcasts, shunned by their neighbours, their children eyed by schoolmates with suspicion.

Shahida's feelings towards the community that ostracised her may be complicated, but her dire forecast for its future is not without foundation. For many foreigners, Bangladesh's vulnerability to climate change has become its biggest claim to fame. It's in a precarious position to start with, being far more densely inhabited than any other large country, with twice Germany's population squeezed into an area smaller than Nepal. It is menaced from the south by cyclones that whirl in from the Bay of Bengal, and in its north by droughts that strike when the monsoon rains prove feeble. Most famously, its heavily populated southern region, formed over millennia by river silt, is exceptionally flat and low lying. If sea level rises by a metre – which, as Koni Steffen told me, now looks like a matter of time – over two thousand square kilometres of Bangladesh could be lost to the sea. But long before that point is reached, the rising ocean will transform still larger areas of the south as salty water percolates ever further inland, with powerful consequences for agriculture and human health. To understand what that could look like, I have come to Gabura.

When Shahida was born in the mid-1960s this community, like most in its region, was sustained by rice farming. But during her early childhood, engineers over the border in India were at work on a huge dam across the Ganges River, which would divert water to clear Calcutta's harbour of silt. As the flow weakened in the rivers that riddle the southwest of Bangladesh, seasonal surges of seawater began to push groundwater salinity to dangerous levels – a problem that then became still worse in recent decades thanks to global warming, as the Indian Ocean began to expand.

Outside the rainy season, Shahida tells me as we walk along a riverside path, her only source of drinking water is a pond near her home. 'It's so salty – but we just have to drink it,' she says. She's convinced that the salty water is behind the health problems of people like her brother-in-law, who's suffered two strokes, and of the local women who seem to be suffering miscarriages with unusual frequency. Recent academic studies back up her suspicion, showing drinking water salinity in this region massively above safe levels, and suggesting a link with elevated rates of miscarriage and infant death.

We halt by a pipe running over the embankment, installed by a shrimp farmer to pump briny river water into the island's interior – an illegal practice that is rife across the region. Shahida spits a red burst of betel nut juice onto the earth. 'They get away with it because they're rich enough to bribe the government people,' she says. 'It's thanks to them that we're poor.'

As the rising salt levels weighed on rice production in southwest

Bangladesh, the national government identified a promising alternative: saltwater shrimp farming. With financial support from the state, shrimp ponds began to proliferate across southern Bangladesh. Today, shrimp is the country's biggest export earner after clothing and textiles. But as this development strategy has reshaped the regional economy, it has had some dark side effects. Shrimp production might be more profitable than cultivating rice, but it requires far more initial investment, while needing much less labour. So countless Bangladeshi farmers, seeing no future in rice, have sold out to the shrimp producers, only to find themselves landless and unemployed. The shrimp bosses themselves are often wealthy outsiders, spending their profits in the national capital, Dhaka, or abroad.

Below us a few islanders are wading through the muddy shallows, dragging blue nets on thin wooden frames, the women in dirt-caked saris, the men in hitched-up sarongs. Nearby sits Mujibur Sardar, shirtless and grey-bearded, using a kitchen bowl to sift through the contents of his net. Like the others he's searching for baby shrimp, tiny as fragments of grass, to sell to the shrimp farmers whose ponds have now almost entirely replaced Gabura's rice fields. He gets a taka – about one US cent – for each one, Mujibur says. If he's lucky he might catch a few hundred. On other days, like Shahida's husband and many other Gabura men no longer able to earn a living from rice, he goes to look for honey in the Sundarbans, where the tigers – their habitat under pressure from the rising sea and the clearance of the forest for shrimp farming – continue to kill dozens of people each year.

It feels like a worrying foretaste of the future of this region, whose salinity problem is set to get progressively worse for decades to come, driven by sea level rise that is already locked in. Bangladesh, according to World Bank predictions, will be the stage for one of this century's great migrations, as millions follow Shahida's four sons in an exodus from the stricken countryside. 'The only people still here,' Shahida says as we resume our walk along the riverbank, 'are the ones who can't get out.'

* * *

Habibullah Mollah wasn't planning to get out. As a teenager in Gabura, watching its more prosperous residents turn growing chunks of the island into shrimp ponds, he wanted to become one of them. His father's rice farm was ailing, but he still had enough savings to loan Habibullah the 200,000 taka – about $2,400 – that the youngster needed to get started. That was sufficient to lease the land, dig the pond, buy the first shrimp and pay a few local men to help him when needed. For a few years things went to plan. Habibullah was earning back his father's money, marvelling at how little work it took to raise shrimp compared with the ceaseless toil of rice cultivation.

Then came 2009's Cyclone Aila, careering in from the Bay of Bengal, unleashing rains of prodigious volume and dragging a mighty wall of seawater towards the Bangladeshi lowlands. When Habibullah emerged from the nearest concrete storm shelter, his farm had vanished beneath the floodwaters that now covered most of the island, his shrimp scattered across a landscape where cow carcasses lay intermingled with bits of household furniture. The better resourced shrimp producers rebuilt their operations in the wake of the disaster, and expanded as they bought up land from rice farmers who now gave up hope after the new injection of salinity into the earth. But Habibullah was broke. He moved to Shyamnagar, the nearest small town, and made a living from casual labour for a few years. But there, too, the local economy was suffering from the decline of rice farming, with the shrimp profits flowing to too few pockets. So Habibullah made his way, along with several million others in the past decade alone, to the swollen slums of Dhaka.

It's a hard life here, he tells me in a tea stall opposite a shed full of cycle rickshaws with brightly coloured awnings over the passenger seats. Habibullah goes there each morning to lease a rickshaw, then sets out onto the city's smoky, potholed streets in search of customers. Having paid for the lease at the day's end, he's left with a few dollars – about $8 on the best days, more like $2 on the worst ones – with which to support his wife and infant daughter. Still, he says, he's earning a lot more here than he could in Shyamnagar, and if he stretches things and skips meals here and there, he can save up to $35 a month. Once he's stashed about $2,500, he says, he'll head back to have another go at shrimp farming. Even, I ask, after what happened last time? 'I'll take my chances,' he grins.

But while many rural migrants come with plans of returning home, most of them are here for good, Abdus Shaheen tells me in his office in downtown Dhaka. When the last national census was taken in 2011, the average square kilometre of the capital housed 43,500 people: the highest figure for any major city in the world. Since then, migrants have continued to pour into Dhaka in their hundreds of thousands, most of them heading for overcrowded slum districts like Bauniabadh, Habibullah's new home.

As the Bangladesh head for WSUP, an international humanitarian group, Abdus is on a mission to tackle the obvious public health menace that arises when millions of people are crammed into informal settlements without access to clean water and sanitation. In Bauniabadh, WSUP's focus has been on building latrines: something that appears nowhere in the list of priorities for the slumlords who build whole complexes of corrugated metal and rent out the rooms for $20 a month. It costs about $2,000 to build a proper community latrine – and, more to the point, it takes up space that could be used to build another income-generating shack.

It seems impossible for charitable groups like WSUP to keep

up with the rush of migrants into Dhaka – a flow that is set to accelerate. Globally, forced migration will be one of the major outcomes of worsening climate change impacts: the World Bank predicts there could be as many as 150 million climate migrants in the next three decades. And Bangladesh is shaping up to be one of the main theatres for this tragedy, it warns, with up to thirteen million citizens set to be displaced – many of them heading to Dhaka. Yet instead of tackling the problem head-on, Abdus warns, the government has been underplaying the dire conditions in the mushrooming slums, keen to tell an upbeat story about rising living standards in the national capital. 'Even the existing problems are not being solved by the government,' Abdus says. 'So what will happen? What is their strategy? What is their plan?'

* * *

Back in the southwestern delta, 25 miles northeast of Gabura, lies the port of Mongla, a rough-edged place with the feel of a frontier town, bisected by a river that can be crossed on wobbly wooden barges for three taka each way. Under searing sunshine I cross the river and jump onto a flat wooden pallet pulled by an electric cycle rickshaw, which halts after a few skull-juddering minutes outside the office of mayor Mohammad Zulfikar Ali.

It's the mayor's birthday and flowers are piling up on his desk alongside sugary rice cakes left by well-wishers. On the wall behind him, as required by government edict, are framed pictures of Prime Minister Sheikh Hasina and her father Mujibur Rahman, who led the country to independence from Pakistan in 1971. The picture of the prime minister is in excellent condition, I notice – unlike the one that Ali is accused of smashing, in a bizarre case making its way through the court system. He's one of many opposition politicians facing criminal charges under the rule of Sheikh Hasina, who secured a third consecutive term six months earlier. Susannah Savage, a young Dhaka-based journalist who covered the election for the *Financial Times*, was one of several reporters targeted by Sheikh Hasina's regime for highlighting the signs of massive fraud and voter intimidation behind that election result. Susannah told me she was detained by Bangladeshi intelligence and deported in handcuffs to the United Kingdom, having suffered physical abuse in custody including a heavy blow to the head. At the time of writing she was awaiting brain surgery in London. It was one of a mounting number of alleged acts of violence against journalists under Sheikh Hasina's rule.

To the wider world, however, the kindly looking Sheikh Hasina has become known as a moral leader in the global conversation on climate change – a role she has played with aplomb, with impassioned calls to action at global conferences and in the op-ed pages of international newspapers. She's in her second term as chair of the Climate Vulnerable Forum, representing forty-eight developing countries highly exposed to climate change, having picked up a UN Champion of the Earth award in New York along the way.

While Sheikh Hasina's government has been praised internationally for spending large sums on climate adaptation measures, it can be hard for opposition politicians like Ali to access the funds they need for their areas, he says. But with the support of a local ruling party lawmaker who's now Sheikh Hasina's deputy environment minister, he's making progress on his vision for Mongla: to build the town into a climate-resilient new home for tens of thousands of southwestern Bangladeshis, offering them a new life in their home region instead of a risky plunge into the distant capital.

'When I took over, there were children fishing in the streets,' says Ali, a softly spoken man with bulldog jowls, who's been Mongla's mayor since 2011. To tackle the flooding from storm surges, he worked with the national government to raise $200 million from foreign agencies to build a 7-mile-long raised embankment. The project also installed sluice gates on the canals that criss-cross this area, to block the ingress of salty water at high tide.

While shoring up the town's defences, Ali is eyeing an economic boom, driven in part by the tussle between Bangladesh's two giant neighbours for influence over the country. China is providing nearly half the funding for a $550 million expansion of Mongla's port. Over 100 acres of the town's southern zone, meanwhile, is being turned into an entirely Indian industrial zone, with operations ranging from agri-processing to chemical production.

Ali expects Mongla's population to double from its current level of 40,000 as the town's economy grows, and as problems continue to mount in the countryside. But he's determined to avoid replicating the situation in Dhaka. Mongla is the first town taking part in a new initiative from ICCCAD, the country's leading climate change research institute. It's aiming to help twenty towns and cities around the country to prepare for a surge in migration. Even if it's too late to save the traditional livelihoods of millions of rural Bangladeshis, the scheme's architects reason, they must be given options other than joining the unsustainable exodus to the Dhaka slums. In partnership with ICCCAD, Ali is building a plan to cater to the anticipated surge of new residents: safe, affordable housing with access to sanitation and clean water, as well as additional storm shelters for the cyclones that will become more fearsome as the ocean rises and warms.

In the afternoon, the mayor takes me for a walk along his prized embankment, an aide trotting alongside him with a parasol. Well paved and lined with slender young trees, it reminds me of riverside parks in more prosperous Asian cities like Seoul, except for the ducks and goats occasionally wandering onto it from adjacent homesteads. Animals are officially banned from the embankment, Ali says. 'We just have to keep telling people again and again until they listen.'

His mood blackens when we reach one of the new sluicegates installed across the town's canals. Instead of holding back the influx of saline water from the high tide, the gate has been lashed open. After the countless accounts I've read and heard of shrimp farmers corruptly obstructing efforts to combat the region's salinity crisis, this looks like another example. A small boy peers curiously from the other side of the sluicegate as its hapless overseer rushes to untie the ropes, under a torrent of fury from the mayor. 'This is crazy,' Ali mutters under his breath.

* * *

In an area of tiny huts and shacks, roamed by goats and chickens and children dancing in sudden downpours, Khairul Mozaffar Montu's house looms tall, shielded by a high perimeter wall and a heavy metal gate that swings open when I arrive for lunch. Five well-dressed men appear and usher me inside the house where I find Montu reclining in a leather armchair, a thickset 72-year-old in a loose white kurta suit.

Montu is the leading shrimp baron of Asassuni, a district to the north of Gabura, and one of the richest men in southwestern Bangladesh. It's four decades since he became a pioneer of the region's shrimp industry, using his earnings from a construction business to dig his first ponds on 50 acres of former rice paddy. Today his farm is thirty times that size, selling over a thousand tonnes of shrimp a year to customers in Europe and Asia; he owns one of the region's most popular holiday resorts with the country's biggest private zoo, and a house near the US embassy in Dhaka, where he spends about half his time.

Montu is familiar with the complaints and allegations levelled against the wealthy shrimp producers who now dominate this region's economy. The most notorious incident of all occurred just a few miles from here, when the leader of a landless women's group was murdered by associates of another large-scale shrimp farmer and politician, at a protest against his expansion plans.

For Montu, however, shrimp farming is simply the only strategy that can offer an economic future for his native region. It's true that his farm employs far fewer people than the land supported when it was used for rice. He only needs to employ one person for every 5 acres of land – and outside the periodic harvesting operations he only even needs them to 'keep an eye on things' and look out for poachers. But he insists it's simplistic to claim that the exodus to Dhaka is driven by shrimp farmers taking over the land. 'The young generation are attracted to the steady routines, the steady salaries,' he says. 'That's why they go to the brick factories and the clothing workshops. Why the hell would they want to stay around here?'

Montu excuses himself to pray, before inviting me upstairs where we are met by a sumptuous spread of fried fish with heaped bowls of rice and chopped vegetables – and lavish servings of shrimp, fat, meaty and cooked to perfection. It's not like climate change isn't causing problems for shrimp farmers too, he observes as we eat. He's convinced that increasingly severe heatwaves are driving outbreaks of disease among the crustaceans. Still, he says, it's at least a somewhat manageable problem, unlike what he sees as a death spiral facing the area's rice industry.

Stuffed from the meal, I waddle to the passenger seat of Montu's new Toyota, which he drives around the corner to the edge of his farm. We stroll between the ponds, surrounded by blue nets to keep out debris, and dotted with electric water mills that oxygenate the water as they churn. An employee pulls out two dozen giant tiger prawns that writhe in furious astonishment, suspended in a finely woven net that lurches and bulges with their thrashing. Montu looks on impassively, the long white shirt of his kurta suit catching in the wind. Give it a month, he says, and he'll get five million taka from that pond. At his signal the shrimp are tossed back to await their fate, and we walk on. 'Whatever happens with climate change, we need to fight back with shrimp farming,' Montu says. 'Nothing else will work.' Around him, the blank grey-green rectangles stretch, on every side, to the horizon.

PART THREE

An Age of Storms

CHAPTER 8

If Venice does not have, in 2019, even a minimum
of protection, the responsibility lies with that political
class that has allowed itself to make money and speculate.
This is the truth and we must say it.

– Luigi di Maio, Italian foreign minister, 14 November 2019

VENICE, ITALY

As the winter night set in over the Palazzo Cavalli, an elegant sixteenth-century pile with a terracotta façade, Alvise Papa remained at his desk on an upper floor, staring through black-rimmed glasses in horror at the data appearing on his monitor. As head of Venice's tide forecasting centre, he was charged with alerting the population to shifts in the level of the lagoon that surrounds them – vital work in a city that, for all its fabled splendour, rests on an agglomeration of marshy islands, most of its land no further above mean sea level than the height of a sheep.

Outside his window the Grand Canal, the most majestic of the city's 150 watery streets, was already swollen well above its normal level, and a high wind was sending the black gondolas straining against their moorings beneath the pale marble of the Rialto Bridge. Since Sunday morning, three days earlier, Alvise had barely left his post, having spotted an unusually strong shift in regional air pressure that promised a rough tidal ride over the coming days. But what he saw now – a sudden shift in wind direction to the southeast, sending a new rush of water towards Venice just at the peak of the astronomical tide – looked set to trigger the worst flood of Alvise's lifetime.

Bypassing the usual protocols in the interests of rapid response, Alvise turned to the alert management system and triggered a code red alarm, the highest notch on the four-point scale. From the famous bell tower in the Piazza San Marco, where once Galileo demonstrated his telescope, an eerie wailing siren rang out as the cold saltwater rose rapidly around its base. Now there was nothing more Alvise could do but watch and wait.

That November night, the tide rose 187 centimetres above the historic mean, covering nearly 90 per cent of the city's land. The water surged into shops, homes and dozens of ancient churches, as motorboats and gondolas drifted over submerged paving stones. 'It was my worst nightmare,' Alvise tells me the following summer in his office, its walls of data-soaked screens incongruous with the building's elegantly faded Renaissance exterior. 'And if it happened once, it can happen again.'

This was the second-highest tide in Venice since records began – and it was just the beginning of the crisis. The most serious alert level – which is triggered by a tide at least 140 centimetres above the historic mean – had been reached only nine times in the whole of the twentieth century. In the past two decades the frequency had increased dramatically, with an event more than once every two years. Now, Alvise was forced to activate his red alert sirens four times in just *five days*.

Born and raised amid Venice's winding canals, Alvise had grown up seeing the winter 'acqua alta' events as an exciting occasional drama, which sent the city children splashing around the squares in gumboots. But now the impact has become far more serious, he says, pulling up a series of charts on a wall-mounted screen to underscore his point.

The charts show the data collected by the city since 1872, when Alvise's forerunners began keeping a daily record of the water level next to the magnificent stone church of Santa Maria della Salute. They're all terrifying. One shows the change in mean sea level, which follows a steeply rising path to a level more than 30 centimetres higher than where it began. Another shows the number of tides exceeding 110 centimetres above the historic mean: there were fifteen of them in the first half-century of the record, compared with twenty-six in 2019 alone. 'That's the real sign of a climate crisis,' Alvise says in the jargon-spattered Italian of his profession, 'when all the parameters are abnormal.'

With sea level set to continue its inexorable rise, the city will become ever more vulnerable to the impact of winter storms, which send surges of water barrelling up the Adriatic Sea, behind the heel and calf of the Italian boot, towards the northwestern shore where Venice lies. According to a study published by ten European academics a few months after our meeting, the frequency of the worst events is set to increase by 160 per cent during this century.

Still, Alvise reflects, even as Venice's magnificent heritage now looks imperilled, it also offers a measure of inspiration. The people of this city, he points out, have a fabled history of ambitious engineering projects ever since, according to legend, they fled the Italian mainland to escape the rampaging hordes of Attila the Hun 1,500 years ago. The first was the creation of the city itself – reinforcing the marshy land to support buildings resting on wooden poles driven into the earth, and creating entire new islands by dumping large quantities of silt taken from the lagoon floor. Then, during the Renaissance, the people of Venice undertook massive works to divert the rivers that flowed into the lagoon, over concerns that the silt they carried would clog the water, depriving Venice of its protection against attackers and killing its maritime economy.

Now, like all Venetians, Alvise is waiting to see what will happen with the city's latest grand project, a multi-billion-euro intervention to protect Venice from the roiling seas, which has generated intense curiosity and raging controversy during an achingly drawn-out gestation. 'It will make a radical difference, if it works,' he says. 'And it really has to work.'

* * *

If Alberto Scotti's plans had gone as he expected, the chaos and destruction of that November flood might have been avoided. I meet him in the foyer of his office on a quiet street in Verona, 60 miles inland from Venice; we stroll over for lunch at a nearby pasta restaurant where the waiters address him as 'Signor Presidente'. A stately figure with huge white eyebrows, Alberto is the boss of Technital, an engineering company that he's built over three decades into an international player with projects including a vast new port city in Iraq. To Venetians, however, he is known as *il padre del Mose* – the father of Mose, a flood defence system whose name, an acronym derived from the Italian Modulo Sperimentale Elettromeccanico, evokes the biblical divider of the seas. Billed as Venice's salvation from watery disaster, it has ended up as one of the most protracted and controversial infrastructure projects in European history.

Plans for a defence system were set in motion more than half a century ago, after a flood in 1966 that swamped Venice for an entire day, devastating homes and businesses and leaving architectural jewels covered in sewage. The process languished for years, with a tortuous series of studies by academics and government officials. In the 1980s a group of Italian companies created a consortium, the Consorzio Venezia Nuova – a name that would later become notorious throughout the country – and won the state contract to develop some kind of flood barrier. But the details remained hazy and the project continued to drift. When Alberto joined Technital in 1987, it was one of a plethora of companies doing nothing very much to establish what the new system would actually look like. Spotting a chance to make his name in the industry, Alberto took some of the academic models that had been developed and drew up a comprehensive design of a sort that had never been built.

Alberto's system would use retractable gates to block the inlets that connect Venice's lagoon with the sea, whenever a surge of dangerous proportions headed northward along the Adriatic. The principle was much the same as that underlying London's Thames Barrier, completed in 1982 to guard against surges sweeping in from the North Sea. But the Venetian authorities had ruled out such a hulking structure, reasoning that it would blight the scenic charm that underpins the city's tourism-focused economy. So Alberto's barrier, building on proposals laid out by Italian academics, would be mostly invisible. On the floor of each inlet, a string of metal gates would lie in a concrete bed. When an incoming surge was detected, operators could activate hydraulic pumps to push compressed air into chambers beneath the gates, which would lift to form a wall, repelling the waves and preventing the lagoon's water level from rising until the surge receded.

In 1989, Alberto went to Rome to submit his plan formally to the minister of public works. For three full days he was grilled by ministry officials and a committee of university professors. 'In the end they were very happy,' he tells me now over a bowl of spaghetti *alla marinara*. It seemed like a triumph. In reality, it was the start of a waiting game that would run for half a lifetime – even as the hardening science of climate change-driven sea level rise made clear that the situation was even more urgent than previously thought. It took five years for Italy's sclerotic bureaucracy to clear Alberto's 'definitive design', then another seven to clear the 'final design', which was approved under the playboy prime minister Silvio Berlusconi in 2002.

At last, Alberto thought, the barrier would now be built. But that clearance merely ushered in a darker new chapter for the project, which would now be hobbled not only by bureaucracy but also by rampant corruption. The Consorzio Venezia Nuova was presiding over a feeding frenzy, doling out contracts to dozens of businesses hungry for a slice of the multi-billion-dollar investment. Although he had no evidence of anything illegal, Alberto began to push back. Apart from the question of how many of these firms had weaselled their way into work for which they were not qualified, he feared the project would descend into chaos as the number of contractors became absurdly inflated. The Consorzio swiftly kicked him out. Cut off from the project to which he had dedicated two decades of his life, Alberto considered early retirement before throwing himself into a new commission to develop the port of Al Faw in southwestern Iraq. From a distance, he watched the gaggle of contractors continue to milk the drifting project with abandon – until it blew up in their faces.

The police swooped on a June day in 2014, arresting a glittering cross-section of the Venetian business-politics nexus, captains of industry along with elected officials including the mayor of Venice. Through inflated invoices and fictitious contracts, a prosecutor told the evening news, Venice's most powerful men had built a sophisticated corruption engine, funnelling huge sums to overseas slush funds. 'I was waiting for this,' Alberto says. 'It was obvious that something like this was happening.' The scandal was a shocking illustration of Italy's weak record on corruption, the worst of any major Western economy according to Transparency International. But it did at least bring an end to Alberto's exile: after the government took direct control of the Mose project, it soon asked him to return to take up a major role.

Progress accelerated markedly after his return, and, by the autumn of 2019, the project was rapidly nearing completion. So when the huge storm surge came rolling in that November night, Alberto's phone began to explode with calls from anxious officials, seeking his go-ahead to raise the gates in a desperate attempt to fend off disaster. But he refused. His team was still working to fix one of a mass of cack-handed mistakes made by the corrupt former management, who had failed to install pipe supports that were critical for the proper functioning of the barrier.

If they raised the gates now, Alberto said, there would be a dangerous level of vibration, which could seriously damage the system and potentially leave the gates jammed upright, blocking the flow of water and shipping in and out of the lagoon for an extended period. Beyond the immediate impact, such a debacle would deal Mose's credibility a blow from which it might never recover. 'I told them, if you oblige me to do this, you will never see me again,'

Alberto says. 'Because we were not ready. We were not ready.' And so the yellow gates stayed flat in their concrete beds as the swollen sea rushed over them towards the palaces and churches of Venice, silhouetted in the night sky.

* * *

Walking through the Basilica di San Marco, Venice's fabled church of gold, I begin to understand how damaging these events can be for such an ancient, fragile city. Months on from the flood, this and other architectural treasures are still undergoing a painful rehabilitation, like an elderly aristocrat struggling to recover her genteel elegance after a severe fall.

Far above my head is one of the world's most dazzling ceilings, blank-faced saints in an otherwise unbroken sea of gold, the work of successive generations of mosaic artists beginning in the eleventh century. In the church's adjoining treasury are hundreds more priceless golden artefacts, looted from Constantinople by Venetian troops who veered off mission during the fourth crusade. Still more stunning is the view from the piazza of the church's exterior, with its Byzantine domed cupolas that still evoke Venice's long-vanished position as one of the world's most powerful cities, the trading hub where western civilisation met that of the east.

I walk beneath the splendour alongside Paolo Gasparotto, the leader of the basilica's restoration operation, amid members of his team kneeling to squirt limestone grout into the cracks in the marble and mosaic floor of the church's nave. Tourists on a balcony high overhead peer down enviously, banished from the main floor until the months-long repair process is complete. In front of an altar on the church's southern flank, Paolo crouches to the ground and rubs some fragments of dust between his fingers – ancient mosaic tiles, pulverised by the invading saltwater. 'We're seeing damage of a kind we've never seen before,' he says.

The basilica owes its existence to the sea, the source of medieval Venice's fabulous trading wealth, of which a big chunk was used to make this one of the world's most famously majestic buildings. But now, the Adriatic's creeping intrusion has left it crumbling from the bottom up.

When the flood hit that winter night, Paolo's team were waiting inside the basilica, having rushed there as soon as they realised what was heading its way. But they could do little more than watch through the shadows as the water invaded the church, oozing up through tiny cracks in the floor. When it finally receded, they set to work bathing the battered surfaces in fresh water, trying to rinse away the salt.

For the real danger, Paolo explains, is not the water itself, but what happens when it evaporates over months from the tiny pores within mosaics and stone surfaces – leaving salt that forms into crystals, which push out against the surrounding material. Paolo's team have started comparing the phenomenon to radiation poisoning in humans. When the floodwaters retreat, a mosaic or section of stone might appear to have suffered little damage – just as the firemen who battled the blaze at the Chernobyl nuclear reactor looked at first unscathed. But in due course, the severity of the damage becomes clear.

We walk over to another ancient treasure: a stone carving of Jesus and two saints, plundered from Constantinople by Venetian crusaders in 1204. In the engraving's lower section, the gold leaf has been flaking away, supplanted by an intense white patina spreading in the marble beneath it. Once that white layer starts to appear, it means the salt is already everywhere, Paolo says, bound to continue cracking and flexing inside the priceless artefacts until they collapse into dust.

If floods like this continue to increase in frequency and severity, the situation could become unmanageable. So Paolo is watching with wary anticipation as the Mose system inches towards completion. 'If it doesn't work, we'll have a big problem,' he says. 'Everything depends on Mose.' But for others in Venice, the barrier is less a promise of a new future than a gigantic symbol of all that has gone wrong in the past.

* * *

Early on a hot Saturday morning, I take a vaporetto – one of the floating buses that ply Venice's waterways – from the main city across to the Lido, a long, slender island on the edge of the lagoon. At a park off a main road I find a few dozen young activists, armed with guitars and hand-scrawled cardboard placards, sitting in a circle listening to Shamir Sartorelli. An affable 23-year-old alternative medicine student with round-rimmed glasses, his yellow Hawaiian shirt hangs open to show a T-shirt bearing a logo painfully familiar to harried London commuters: the jagged encircled sand timer of Extinction Rebellion.

His pep talk completed, I walk with Shamir towards the coastal road where he's leading today's protest against plans to develop a virgin stretch of the Lido's eastern coastline into a leisure resort, destroying an expanse of dunes that serve as a natural defence for the lagoon system, cushioning the impact of storm surges. For these young activists, it's a classic example of the missteps that have got Venice, and the world as a whole, into such a fix: well-connected businessmen degrading a community's environment and security, for the sake of commercial gain and another luxury weekend option for the rich.

Like many green-minded young people across Europe, Shamir was intrigued by the sudden emergence, in late 2018, of a new environmental movement that threw central London into gridlock with a series of roadblocks manned by thousands of protesters, demanding radical action against climate change. 'We . . . declare ourselves in rebellion against our government and the corrupted, inept institutions that threaten our future,' Extinction Rebellion said in its founding manifesto. Climate change, its leaders declared, was evidence of the failure of modern democracy.

Extinction Rebellion Venezia was formed in the late summer of 2019, just a few months before that winter's disastrous storm. The leaders decided from the start, Shamir says, that the confrontational approach of the British mother movement would not translate well to Venice. Massive disruption might be the only way to get noticed in a city as big as London, they reasoned, but in a small, genteel place like Venice it would probably just create pointless ill will. Instead, the flood gave the group a chance to make a positive impression on their fellow Venetians, by organising teams to help families start putting their damaged homes back in order. And, of course, it meant they had to spend a lot less effort in persuading people of the dangerous effects of climate change.

Here at the Lido, they've continued their relatively gentle approach: blocking only half the road through a staggered series of stations where activists sit on the ground with placards, forcing drivers to navigate around them while the protesters try to thrust educational leaflets through their windows. The mood is more summer festival than angry riot: as Shamir and I talk at the roadside, activists dance around us to a comrade's Venetian-accented rendition of 'Redemption Song'.

On only one occasion do tempers flare, when a dog walker in tiny swimming shorts below an impressive chest rug starts haranguing a skinny, nervous-looking activist about his group's interference in this community's local affairs.

'Go on, tell me! Where are you from?' the man shouts.

'Mestre,' the protester replies miserably, referring to an industrial district on the other side of the lagoon.

'Pah!' the man yells, leading his dog away in triumph.

'Most people don't seem to care so much,' Shamir says, returning from a doomed attempt to chase after the dog walker and win him over. 'They're used to the fact that this is how things are. They feel powerless, so they just don't care anymore.'

To a certain extent, he goes on, this apathy is understandable, when one considers the dire track record of Italian politicians from across the political spectrum. For Shamir, there's no better illustration of this than the Mose corruption scandal, which unfolded over more than a decade under the supposed oversight of local and national politicians from right and left alike. It's yet another piece of evidence in the case for one of Extinction Rebellion's central demands: the creation of Citizens' Assemblies, their members selected at random like those of a criminal jury, whose rulings would determine major government policy decisions.

The authorities could have focused on preserving Venice's natural protections, Shamir says. They could have given more consideration to relatively low-cost interventions, such as a plan devised by scientists at the nearby University of Padua to inject water into depleted aquifers beneath the city, gradually raising its ground level. Instead, they went for the most expensive, spectacular option, one which offered the most lucrative prospects for Italian engineering firms, but which – according to environmentalists – threatens to pollute the lagoon and disrupt its exchange of water with the sea. Like everyone else at the protest, he pours scorn on the notion that Mose will work as planned, even after the billions that have now been ploughed into it. 'But now that they've spent so much time and money on it,' he adds, 'they can't turn back without losing face.'

* * *

A few days before leaving Venice, I join Alberto, Italian prime minister Giuseppe Conte and a few dozen other officials and journalists on a small artificial island, in the middle of the lagoon's widest inlet, which serves as Mose's command centre. It has the look of a military fortification to fend off a seaborne invasion, suffused with mammoth concrete slabs roamed by soldiers drafted in to work on the communication system ahead of today's big event: the system's first full-scale test.

Over by the docks adjoining the Piazza San Marco, a small naval confrontation is under way, with police and coastguard vessels scrambling to block a dozen boatloads of environmental protesters from disrupting the event. It's right that they should make their voices heard after the 'absolutely deplorable episodes' of corruption, Conte, a smooth-talking former law professor, tells the dignitaries on the island. His two-year-old government didn't design this structure, he reminds his audience, but after all the billions already sunk into it, the only sensible thing to do is to get it finished and 'hope it works'. It's a strikingly half-hearted endorsement, a reminder of how politically toxic Alberto's project has become over the long years of delay and scandal.

The procession moves on to a viewpoint on the edge of the island, overlooking one side of the inlet, where a shaven-headed priest in a heavy white cassock is waiting to bless the barrier with holy water. Just as he shakes his tiny metal sprinkler at the sea, the first of the yellow barriers pokes its nose above the surface. It rises at a stately, almost languid pace, then comes to rest at its final position, pointing out to sea at a forty-five-degree angle. For some moments it stands there in awkward isolation, and I wonder whether the day could turn into a hideous fiasco. But then I see another rising on the other side of the passage, followed by a third, and a video screen to one side shows the same scene taking shape at each of the lagoon's other three openings. Within half an hour, all seventy-eight of the gates are upright. For the first time in history, the Venetian lagoon is cut off from the sea.

Cruising across the still surface of the inlet on a small yacht emblazoned with the winged lion of Venice, Alberto stands among the assembled political leaders to watch his vision made real after thirty-three years of frustration. He's too savvy to think this moment will silence Mose's many doubters. The system may have worked on this single calm summer's day, they'll ask, but how will it fare when deployed in rough winter waters, multiple times in quick succession? Will the barrier save this fabled city for future generations – or itself become an obsolete historical relic, neutered by increasingly destructive storms powering through a rising ocean? Today, Alberto can enjoy his moment of vindication. But the answers to those questions will come soon enough.

CHAPTER 9

To date, most human rights bodies have barely begun to grapple with what climate change portends for human rights.

– Report to the UN Human Rights Council by Philip Alston, special rapporteur on extreme poverty and human rights, 17 July 2019

MANILA, PHILIPPINES

By the entrance to a soaring glass office tower, Joanna Sustento remains silent as the police line up before her, more than thirty-strong in pressed navy blue slacks and short-sleeved shirts, guns and handcuffs at their waists. Approaching from the nearby police station in Manila's central business district, some have their jaws set in well-practised Robocop scowls; others look coyly uneasy to be part of such an overwhelming force deployed against a lone, slightly diminutive young woman.

Habagat, the southwest monsoon, is in its final weeks, and a weak drizzle drifts down upon Joanna's red windcheater and black side-parted bob, and on the placard thrust high above her head. The police move in swiftly, a female officer guiding the unresisting protester towards a waiting black car while others whisk the offending sign out of sight. It bears an image taken by a drone in Joanna's home city of Tacloban, on the fifth anniversary of 2013's Typhoon Haiyan, which killed more than six thousand in the city and its surrounding region. Citizens gathered at a memorial site and arrayed hundreds of candles that cast a glow on their faces and formed a message clearly visible from the air: *CLIMATE JUSTICE.*

When I meet Joanna for tea the next day, she evinces no anger towards the police, who let her go on her way after half an hour at the station, and said they were merely trying to avoid a scene. This was good treatment by the standards of law enforcement under the strongman president Rodrigo Duterte, who won the moniker 'Duterte Harry' for his boasts of summarily executing suspected criminals when a city mayor, and as president has overseen the street shootings of several thousand alleged felons.

Joanna's fury remains locked on the target of her protest: Shell, the corporation whose Philippine headquarters sits on the forty-first floor of that Manila tower. For more than a century it has been a giant of the global oil industry, one of the central drivers of the soaring carbon emissions distorting the global climate – and therefore, in Joanna's eyes, a culprit behind the typhoon that stole her family.

On the day Haiyan hit, she wasn't immediately alarmed to wake up with rain gushing in through her bedroom window. The weather service had forecast a severe storm – a regular fact of life in the central region of the Philippine archipelago. In their sturdy, comfortable bungalow on the southern side of Tacloban, even as it vibrated from a wind that was starting to bend the trees outside, Joanna's family gathered for breakfast as normal. Her father Cesar who had fallen in love with her mother Thelma when, as a security official under the dictator Ferdinand Marcos, he was ordered to spy on the pretty young political activist. Her chubby elder brother Julius, and oldest sibling Jonas, who had quit his rock band in Manila to move home and open a tattoo parlour. With Jonas had come his wife Geo and their son Tarin, now aged 3, who dissolved any family tensions as he toddled through the house singing to himself.

Suddenly water was covering the living-room floor. Joanna rushed to her bedroom to grab some clothes and an iPad, but while she was in there the water rose to her chest, and Jonas was yelling for everyone to get out of the house. The tremendous force of the storm, with winds that would reach a peak force of nearly 200 miles per hour, had created a colossal mound of water that was now powering inland through the narrow bay to Tacloban's east. The aim was no longer to salvage possessions, but to stay alive.

The family pushed past floating furniture to their front yard, where they clung to the house's window grilles to avoid being swept away by the rampaging sea. Julius was the first to go, after trying to retrieve an icebox that could have offered a floating refuge for little Tarin.

Next Geo lost her grip and hurtled away from the house along with her baby son, dwarfed by the adult lifejacket his uncle had strapped around him. Jonas thrashed towards his wife and child while Joanna turned her attention to her parents, clinging to the edge of the roof as the sea rose around them. When she looked back, Jonas and his young family were gone.

Joanna seized a passing log and helped her parents grab onto it. The current pulled them inland, through streets turned to furious rapids. After some minutes her father lost hold of the log and sank beneath the roiling water, resurfacing briefly before vanishing for good. As they were swept on through the submerged city, Thelma fell unconscious. Badly weakened after what was now an hour in the water, Joanna held on to her motionless mother as long as she could.

When the waves at last receded, Joanna was alone atop a water tank, watching a ruined city drift into view, its colours drained, its buildings flattened, its surviving citizens slowly descending from trees and lamp posts, stumbling silently, mechanically, through streets carpeted with the intermingled cadavers of people, pigs and cats.

In the weeks and years that followed, Joanna's shock and grief turned to rage. She had found Julius wandering in the streets, injured but alive, and together they had retrieved the bodies of their mother and brother and sister-in-law. Their father and nephew were never found. Later, financial help was promised to bereaved survivors, and Joanna lined up for hours among fellow citizens waiting for assistance. She was appalled by their resignation to the fact that they would receive far less than they were entitled to, so that corrupt officials could take a cut.

Her anger grew as she learned about scientific assessments of how global warming has been increasing the force of storms like Haiyan, by driving up sea surface temperature and atmospheric moisture levels. It was fuelled every time she heard people talk about these storms as acts of God; when she heard defences of oil companies on the grounds that they had contributed a few million dollars to reconstruction and relief after Haiyan. It rose each time she heard experts urging communities like hers to adapt to a hostile new climate, as multinational groups raked in billions while propelling them towards that fate.

'We can't just keep adapting,' Joanna tells me in a cramped Manila tea shop, the words delivered evenly and with steely precision. 'We need to fight.'

And as she speaks, in a building on the other side of town, a small team of public officials are working to complete a document that could prove a powerful new weapon in her struggle.

* * *

As a student activist in the 1980s against Ferdinand Marcos's brutal kleptocratic regime, Roberto Cadiz used to fall asleep each night wondering if he would be shaken awake by the secret police. But the activists' work paid off. In the wake of his brutal attempt to steal the 1986 general election, huge crowds flooded the streets of Manila to demand Marcos's resignation. The fallen strongman and his glamorous wife Imelda – having looted the country of billions during twenty years in the presidential palace – fled to Guam within days.

Cadiz completed his legal training and built a lucrative career as one of Manila's leading lawyers. But in 2015 he went back to his roots, accepting a seven-year appointment as one of five commissioners running the country's Commission on Human Rights. He had barely settled into his new office, in the commission's squat headquarters off a major highway, when a curious document landed on his desk – a petition from local activists calling for him to start an investigation into the world's leading fossil fuel companies.

These businesses, the thirty-two-page petition argued, had cynically worked to deepen the world's addiction to fossil fuels in full knowledge of the devastation that would ensue. Some had actively sought to undermine public understanding of that danger, by funding mendacious assaults by cranks and charlatans against the hardening scientific consensus on anthropogenic climate change. Meanwhile, communities across the Philippines, and around the world, were already being swept by consequences that would only become more severe. This was a matter of human rights, said the petition, and it was time to start treating it as such – especially in the Philippines, one of the countries most exposed to the impacts of climate change. The commission must launch an inquiry.

Cadiz's first instinct was to throw the petition in the trash. When he had joined the commission, a body set up in response to the Marcos regime's abominations, his ambit had seemed pretty clear. People killed and wounded by guns and batons. Torture. Rigged elections. Oppression of minority groups. If a business was knowingly poisoning the water supply of the people living around a factory, that might make the grade. But to pursue a company for selling petrol to American or European drivers, in connection with typhoons in the Philippines, a country where that business might not even operate? This was far outside Cadiz's comfort zone.

Over the following days, however, the questions raised by the activists gnawed deep grooves in his brain. If a fraction of such harm were knowingly caused in any other context, the human rights case would be compelling. Lawsuits had already started to crop up in other countries, aiming to force courts and governments to treat climate change as one of the greatest human rights violations in world history. In the US, twenty-one children had sued the national government, accusing it of harming their right to 'life and liberty' by failing to take serious action on climate change. In Peru, farmer Saúl Luciano Lliuya was launching legal action against German power company RWE – Europe's biggest carbon emitter – seeking to make it pay its share of the cost of protecting his village from a dangerous lake created by a fast-melting glacier.

So far, those cases were struggling to make headway, with courts deeply sceptical about the idea that 'climate justice' fell within their remit. But if a new way of thinking began to develop in the global legal community, viewing climate change through the prism of fundamental human rights, this might change. Instead of dumping the direst costs of their products on the rest of the planet, the most powerful fossil fuel companies might be held to account. Cadiz took on the case.

When we meet in his red-walled, windowless office at the CHR headquarters, Cadiz is in the thick of his investigation, now persuaded that climate change presents the most urgent human rights problem of our time. He wants to spread this message, not only in the Philippines, but throughout the world. 'It's easier for people to understand when their rights are violated by a dictatorship,' Cadiz says. 'Climate change is a bit more challenging.'

Especially, I suggest, since most of us are complicit. The fossil fuel companies are on one side of the transactions that send carbon dioxide swimming into our atmosphere. But on the other side are their customers – especially the world's richest few hundred million people – who are the end users of their energy, relying on it to heat our homes and freeze our food, produce our clothes and gadgets, transport us to work and away on holiday.

'But do you really have a choice, Simon?' Cadiz replies. In 1988, when NASA scientist James Hansen put climate change in the political spotlight with a groundbreaking address to the US Senate, optimists hoped that huge, concerted investment in green power would provide the world with a vastly cleaner set of energy options, enabling dramatic reductions in carbon emissions. Yet since Hansen's speech, more carbon has been released from fossil fuel combustion than had been generated in world history up to that point. This, Cadiz believes, is due in large part to the fossil fuel companies – both their slowness to embrace renewable energy, and the industry's long-running efforts to undermine the hardening climate science.

Crucially, this strategy enabled them to amass vast profits, while communities and governments – disproportionately in poorer parts of the world – were forced to cover the costs of the climate change that they were fuelling.

In the year before Joanna's protest at its Manila office, Shell paid out more money to its shareholders than any other company in the world: $20 billion, comfortably beating second-placed Apple. Its chief executive Ben van Beurden earned over $62,000 a day. Such fantastic rewards were possible only because the full costs of Shell's products were being shouldered by others, who would continue to bear them – along with people yet unborn – far into the future.

No national human rights body had ever taken on a case like this before, and Cadiz had to think carefully about how to organise the hearings. He decided three kinds of voices should be heard: survivors of climate change-linked disasters, who could speak to the human impacts of these events; scientific experts, who could lay out the complex causal chains behind them; and representatives of the fossil fuel companies, who would be given a chance to defend their record.

The hearings were held over nine months in Manila, New York and London. Joanna Sustento was among the dozens of disaster victims who took part; James Hansen among the many scientists. But not a single voice was heard from Cadiz's third category. Shell and other fossil fuel companies deployed lawyers in an attempt to shut down the inquiry before it began; when that failed, they refused to testify, unwilling to give any additional legitimacy to the process. Cadiz's seasoned eye detected an inner turmoil among his former bar colleagues now representing Big Oil. Their awkwardness reminded him of lawyers he had seen defending criminals they knew to be guilty.

Cadiz is braced for a withering response from the fossil fuel industry when he delivers his findings. But he hopes his commission can play a role in shifting the terms of this debate. The right to life, the right to food, the right to water, the right to health – all are being grievously infringed among the world's most vulnerable communities by the fallout from an industry generating billions of dollars in profit every month.

Cadiz knows it's unrealistic to expect the fossil fuel companies to cover the full cost of their historical liability – the resultant flood of bankruptcies would plunge the world economy into crisis. But if his report can help pave the way towards a multilateral treaty, enshrining in international law the human rights implications of climate change, then that in turn would force courts to take far more seriously the lawsuits that continue to be filed in jurisdictions around the world.

If just one of those suits were successful, it could transform the dynamics of the fossil fuel industry, as the costs of climate change come to bear on the energy giants' swollen balance sheets. There's an obvious precedent in the form of the late-twentieth-century lawsuits against the US tobacco industry, which ridiculed the cases before finally agreeing to pay out hundreds of billions of dollars. 'The first successful case will open the floodgates,' Cadiz says. 'And we can help open those floodgates.'

The Philippines is beginning to make its mark on the global conversation around climate justice. But far from the spotlight, in Joanna's home city of Tacloban, that conversation has taken on a bitter, deeply local form.

* * *

On her final day in this house, now stripped of roof and walls, Lucia Milado is having a strikingly busy last few hours of business, selling snacks and cigarettes from her little home store to the men who have come to rip it down. Wearing thick gloves and bandannas that filter the great clouds of dust they create, the workers move haphazardly from one flimsy home to another, pulling and tearing and hammering at the plywood walls and thin metal roofs. They yell to each other over the clanging sounds of rushed destruction, mixed with the cries of cockerels sounding their dawn chorus in mid-afternoon, as if in solidarity with the dazed, disoriented residents of the disintegrating Barangay 48.

Like all her neighbours, Lucia has been given just days to vacate the house where she has lived for twenty-nine years, before its demolition and her forced departure to a new settlement 6 miles to the north. For several years, the Tacloban city government has been trying to persuade residents of the city's coastal slums to move to the new developments built for them by the state in Haiyan's wake. Much further inland and on higher ground, the residents of districts like Barangay 48 will be safer there, the government says. But the new homes are small and typically without electricity connection, and access to water is problematic, with boreholes often running dry. The distance from the city and the coast is wearisome for people who have to work long hours in downtown Tacloban, or as dock workers or fishermen. It means time-consuming round trips in jeepneys – the brightly painted public carriages that ply the roads of the urban Philippines, old Japanese trucks equipped with rows of benches along which passengers, haunches pressed together, pass their coins from hand to hand towards the driver. And the cost of that commute is painful for people earning five or six dollars a day.

So, despite the devastation of 2013, Lucia and her neighbours chose to rebuild their houses by the sea, and take their chances against the typhoons. Now they've exhausted the patience of the city government under Mayor Alfred Romualdez – the stocky but photogenic scion of a family that was already the region's most powerful when its most eligible daughter, the young Imelda Romualdez, married Ferdinand Marcos.

The mayor's backers have been keen to distance him from the corruption and violence that occurred when his aunt Imelda was in the presidential palace, amassing thousands of pairs of shoes and reputedly becoming the single biggest customer for the global jewellery industry. He was 3 years old when Marcos took power; still only 24 when the regime tumbled. But now the mayor's opponents accuse him of showing an authoritarian ruthlessness worthy of his late uncle – all under the banner of climate resilience.

Lucia is stout and well into late middle age, her hair a tinted auburn, her blue eyes warm but her shoulders slumped with defeat beneath her modest checked dress. On the floor of her devastated home lies a melange of debris and household items – a broken pair of over-ear headphones, a box of Turkish Delight. In one corner a cracked toilet bowl, exposed obscenely by shattered walls. And smiling out from a grubby flyer in the rubble, the handsome face of Yedda Romualdez, a former beauty queen who married the mayor's cousin and is now the latest member of the clan to run for parliament.

Lucia is miserable at the thought of moving to the hills. It will be hard to earn a living from a small snack shop when everyone else up there seems to be trying the same thing, given the lack of other options. 'But we have no choice,' she says. Like most inhabitants of informal settlements in the Philippines, she has no formal title to the land where she lives. It belongs to the government, which is legally entitled to assert its control. Her son Marlon, a scowling presence in an oversized black basketball singlet, has a more defiant take. 'What they are doing here,' he says, 'is a human rights violation.'

Later I sit in the kitchen of Fernando Magdua, the elected chairman of the barangay, a tough fisherman with a shaved head, surrounded by his sons and grandchildren under the thin light of a naked bulb. Weeks after Haiyan, his twin sons Ronnie and Ronrey rallied the neighbourhood to build a huge Christmas star bearing the Philippines flag, in what was celebrated by media across the country as proof of their city's indomitable spirit. But now all sixteen residents of this house, one of the largest in the district, are being forced to leave, the family scattered between several smaller homes in the new northern settlements.

Fernando is the only one of the district chairmen, he tells me, who tried to resist Mayor Romualdez's plans, arguing with him that the people should be given a choice on whether to continue living in this dangerous area. There are good reasons why fishermen do not normally live up in the hills. 'Typhoons are not a permanent thing. We have to think about our livelihoods,' Fernando says. The mayor was unmoved, telling Fernando that this was a policy that came right from the top. The clearances, he told Fernando, were in line with the instructions of President Duterte – something which Fernando thought probably explained the extraordinary haste with which they were being carried out.

I walk out of Fernando's house into the rubble as night falls over the devastated barangay. It has been returned almost exactly to its state immediately after Haiyan. A couple of youths, wandering amid the ruins, wave to me and smirk in defiant black humour. 'Typhoon Romualdez!' one of them yells.

* * *

I meet the mayor on a Friday night at Casa Luna, a restaurant and music venue in an imposing whitewashed colonial building on one of Tacloban's central thoroughfares, down which flows a tangled stream of jeepneys and motorised rickshaws. Alfred Romualdez is a tall, jovial figure with an almost cartoonishly expressive face and a voice to match, frequently rising up to a near-falsetto for emphasis. He almost ended up among Haiyan's victims: he found refuge in the rafters of a resort building that he owned next to his home, emerging to find that his family had survived only by clinging to a roof, with one of his daughters swept away and left hanging onto a lamp post.

He became the global face of a stricken Tacloban after the typhoon, launching a row with then president Benigno Aquino over the national government's supposedly heel-dragging response – a dramatic example of the political tensions that have complicated all kinds of climate responses around the world. Romualdez is still seething as he lays out allegations that support from Manila was held back by old grudges – presumably linked to the murder of Aquino's politician father, which the family always blamed on the Marcos regime.

But having won attention by loudly standing up for the victims of Haiyan, Romualdez is now under scrutiny for his own treatment of them. He tells me he faced sceptical questioning, even before the demolitions got under way, from human rights officials at the UN concerned about how the migrants would adjust to new locations far out of town. He responded by hailing the resourcefulness of his city's people. Fishermen often face periods when they cannot set to sea, Romualdez says; instead they turn to carpentry or work as waiters. 'They multitask!' he remembers telling the UN team. 'Don't tell me I don't know what my people are doing.'

It's easy to pick holes in the plan, he goes on, if you haven't had to deal with the appalling carnage of a typhoon that ripped through coastal shanties. 'We lost five hundred children,' Romualdez says, raising his voice over the jaunty music. He personally supervised the burial of 2,000 bodies. 'I don't want to do that anymore.'

Many of the houses, he allows, have been shoddily built by the national housing authority. Many more still lack electricity and water access. But they will still keep people far safer from the storms whose violence is an increasingly clear threat. The land those people vacate can then be used for other purposes: a bike path, parks – and commercial development. The building ban applies only to dwellings. There is no embargo on building shops or offices, which the government reasons are easier to evacuate and mostly empty at night.

A few days earlier at a coffee shop in central Tacloban, one of Romualdez's fiercest critics told me this was the plan all along – to clear the poor from what is, despite the typhoon risks, still prime commercial real estate. Climate change was just an excuse for the mayor to pursue a glitzy development plan, clearing the poor far outside the city limits and making way for shiny commercial buildings that could help Tacloban keep up with fast-developing cities elsewhere in the Philippines. A classic case, in short, of what the Canadian writer Naomi Klein has dubbed 'disaster capitalism'.

That isn't it at all, Romualdez says. He insists that he's doing what's best for the city's vulnerable people – even if they might not yet realise it. 'Being a mayor is like being the father of a city. In the beginning, everyone opposes change,' he says. 'But if we have a new normal, we need new thinking.'

* * *

Romualdez refused to follow the example of mayors in other typhoon-hit communities, he told me, by allowing the dead to be buried ad hoc all over the city, which would have left residents stumbling on bones for decades to come. Instead he took over a large tract of land at the back of a private cemetery on the fringe of Tacloban, where workers deposited the corpses of over two thousand citizens in yawning pits.

'The bodies were dumped like rocks,' Joanna says, as we walk past the tombs of Tacloban's elite dead, resting in tenderly maintained graves at the front of the cemetery. In the distance looms a huge cross erected in memory of the Haiyan victims, pressing into a sky of bulging grey clouds that promise imminent respite from the sticky heat. In front of the memorial runs some kind of wild, luxuriant meadow, tall tangles of rich tropical green. As we draw closer, I begin to make out the many hundreds of white crosses beneath the rampant vegetation, each commemorating one of those who lie in the mass grave below. It is scarcely recognisable from the impressively groomed site through which Romualdez escorted Al Gore two years earlier, a central scene in the second big climate film released by the former US vice president.

We wander up to the lofty memorial, by which dozens of slabs of marbled black stone present a dizzying list of the thousands of dead, organised by forename. Three Bernaditas, five Nimfas, twenty-nine Marks. Joanna treads slowly past the names, pointing to those of lost friends and relatives as though picking out faces in a crowd.

Near us loiter half a dozen small boys who have wandered over from the beat-up houses behind the memorial, from which waft the insistent rhythms of Filipino love songs. They'll get a chance to earn some pocket money in a few weeks, they say, when the government rushes to clean up the derelict graveyard in time for the sixth-anniversary commemoration. President Duterte is expected to attend, and stand with Mayor Romualdez in remembrance of the disaster, having inspected progress on the dramatically accelerated relocation drive.

Joanna was there in Barangay 48 the day before, wandering amid the ruins of family houses, past residents rushing with bags of belongings to the trucks that would take them to the new hillside settlements. She struggled to sleep that night. Partly it was the devastation, so uncannily reminiscent of that wrought by Haiyan. But it was also the words that kept running through her mind, which we heard again and again from people hurriedly fleeing their homes. *Waray naman kami pagpipilian*. 'We have no choice' – the words of people caught in a web of interlocking forces, facing the ultimate consequences of decisions made in distant boardrooms which had rippled through, in strange and unpredictable ways, to some of the world's most vulnerable communities. Even as she works towards it as an activist, Joanna knows that true climate justice remains a distant prospect. But as the full force of climate change becomes clear, corporate titans, too, are facing an era of upheaval.

CHAPTER 10

Insurance and economic losses caused by climate-related events are likely to start trending upwards as a share of GDP . . . Disregarding the implications of climate change can generate significant risks for the financial sector.

– Speech in London by Christine Lagarde, president of the European Central Bank, 27 February 2020

MUNICH, GERMANY

On a crisp Bavarian morning I walk over from the Giselastraße U-Bahn station towards the loping mannequin, smooth white limbs frozen mid-stride, that guards the entrance to Munich Re's huge but otherwise unobtrusive headquarters. Huge but unobtrusive is a decent three-word description of Munich Re itself. Since its 1880 foundation, it has grown quietly into a global financial giant, with a scale – its annual turnover is about $70 billion – to match that of its full name, Münchener Rückversicherungs-Gesellschaft.

Munich Re stands at the apex of an enormous system built to soften the sharp edges of life in a turbulent and unpredictable world. The growth of the modern insurance industry has enabled people and businesses around the globe – albeit still disproportionately in the rich countries – to protect themselves against pretty much any imaginable calamity. That insurers are able to take on so much risk, at prices affordable to their customers, is largely because they are very often not bearing the risk themselves. While selling you protection against threats to your health or property, your insurer is buying its own protection from so-called reinsurers, hedging against the risk that you and thousands of your compatriots might suddenly fall prey to an epidemic or natural disaster. For most of the past few decades, Munich Re has been the biggest beast in reinsurance, fending off competition from Swiss Re in nearby Zurich. For huge swathes of the world insurance industry, the buck stops in Munich.

For a company with such a systemically important role in modern civilisation, Munich Re has generally kept a remarkably low profile. (It did inadvertently hit the tabloid press in quite spectacular style in 2011 when news leaked of an orgy thrown for a hundred members of its sales division, complete with colour-coded armbands to indicate which of the female guests were reserved for the top-performing salesmen.) In recent years, however, Munich Re has dropped its aversion to the limelight to become one of the most prominent corporate voices in the global conversation on climate change. The voice, to be more specific, is that of Ernst Rauch, the company's energetic global climate head, whom I find at a high table stacked with biscuits after a long walk through whitewashed tunnels in the building's bowels.

Wincing behind his rimless spectacles at how starkly it dates him, Ernst describes the expensive computer that he used to develop Munich Re's first catastrophe model in 1988: a cutting-edge IBM XT 286 with a whole 640,000 bytes of RAM, or 0.016 per cent of the computing power of the latest iPhone. It was Munich Re's second computer – the first having gone to the CEO, who barely used it. Still, Ernst's assignment marked an escalation of one of the most forward-thinking responses to climate change of any major company. In 1974, Munich Re had set up its first operation to analyse evolving patterns in natural disasters, under a meteorologist named Gerhard Berz. His team used methods that seem absurdly primitive by today's standards, collecting newspaper clippings on disasters around the world, and compiling the findings in handwritten ledgers. But Berz had hit – far earlier than most in this industry – on an insight that would change the sector forever. Natural disasters might no longer be entirely natural. Human industrial activity, Berz's gut feeling told him, could be moving catastrophe patterns outside their natural bounds.

A few years after Ernst joined Berz's team, an Atlantic storm sent shockwaves through the whole global industry. Sweeping in from beyond the Bahamas in August 1992, Hurricane Andrew smashed into the suburbs near Florida's southeastern tip with winds that reduced entire neighbourhoods to rubble, leaving 250,000 people homeless. At $25 billion, the damage bill was the highest of any disaster in history. Most of that, dazed Floridians realised with relief, was covered by their insurers.

'It was a really epic moment for the industry,' Ernst recalls. Sixteen insurance groups went bankrupt in Andrew's wake, and the survivors realised they needed to start putting far more work into managing their disaster exposure or risk meeting the same fate. It was a wake-up moment that transformed the sector. The property insurance business had been built on the notion that the past was a reliable guide to the future. Using historical loss statistics, underwriters calculated the odds of a large payout, and the appropriate premium to charge for taking on that risk. But Andrew's losses had no precedent in the historical record – and it could easily have been much, much worse. Had Andrew passed just 20 miles further north, the eye of the storm would have flattened downtown Miami. Belatedly, the whole global insurance industry began to follow Munich Re's lead. It was time to start investing in technology, fast.

Today, every major insurer has a team using high-powered software and the latest atmospheric research to build constantly updated models showing the risks of major disasters all over the world. Ernst now leads a Munich Re team of thirty analysts on three continents, running simulations to assess the risks to the company's finances – over both the next few years and the next few decades. Their work has provided grim confirmation of Gerhard Berz's hunch half a century ago. Ernst's models now show a starkly increased danger of wildfires in places like California and Australia, reflecting the disastrous blazes seen there after protracted dry periods in recent years. And in the western Atlantic – by far the industry's single biggest hotspot – the threat of severe hurricanes is ratcheting up. 'The very strongest ones really have become stronger,' Ernst says. Starting in 2016, the Atlantic was hit by Category 5 storms – the strongest type, once a rare event – for a record five years running. In 2017 alone, the Caribbean and southeastern US were pummelled by three massive hurricanes in a single month, with a total damage bill twelve times higher than Andrew's.

Despite the unprecedented run of disasters in recent years, insurers have stayed in business. Now awakened to the scale of the danger, the companies have amassed far larger loss reserves than they had before Andrew, and taken a much more careful approach to the protection they offer. It's that latter point, rather than the prospect of bankrupt insurers, that the world might want to start worrying about, Ernst says. As climate risks increase, the price of insurance for some sorts of property – principally in coastal areas, and near forests – could reach unaffordable levels. 'It could undermine our business model in the long run,' he says. 'But as a private sector company, our price tag has to be in line with the risk.'

* * *

Six weeks after sailing west from Plymouth in 1609, the theatre investor and sometime poet William Strachey was caught in the clutches of a storm whose fury far surpassed anything he and his shipmates had seen before. His account of the event – widely held to be the inspiration for Shakespeare's *The Tempest* – remains one of the most evocative descriptions of an Atlantic hurricane, one that 'did beat all light from heaven, which like an hell of darkness turned black upon us'.

For three days the 150 passengers and crew fought to stay alive, endlessly bailing out the ship and throwing overboard its cannon and barrels of beer and wine, while 'the waters like whole rivers did flood in the air'. On the fourth day, as hopes of survival dwindled on the badly leaking vessel, an island came into view, and the exhausted sailors managed to run the ship aground on its rocky shore. 'We found it to be the dangerous and dreaded island, or rather islands, of the Bermuda,' Strachey wrote, 'called commonly "the Devil's Islands".'

Previously regarded as a fearsome shipping hazard, Bermuda turned out to be an excellent spot for a colony, proving a valuable asset for British naval strategy. And four centuries after the storm that opened Bermuda's history, hurricanes are at the centre of its latest chapter. With its minimal taxes and friendly regulators, this tiny territory of 64,000 people has become the hub for one of the global financial sector's fastest-growing, and most eye-catchingly named, asset classes: catastrophe bonds.

It's a field that has attracted scant attention, even in the financial press. But since its inception in the late 1990s, the market in catastrophe bonds and related 'insurance-linked securities', or ILS, has grown from nothing to nearly $100 billion. And that expansion is set to continue, as managers of the enormous pool of global pension savings awaken to the charms of these unconventional investments. If you're a Coca-Cola employee in the US, or a nurse in North Yorkshire, or a railway worker in Switzerland, or any citizen of New Zealand, then a chunk of your retirement nest egg is already sitting in this market. And so, geekily intrigued, I find myself in Bermuda's little capital of Hamilton, a faintly surreal hybrid of quaint English market town and slick financial district, fringed by yachts bobbing in milky turquoise shallows.

The basic idea of catastrophe bonds is to enable market investors to wager against the likelihood of natural disasters, typically with an insurance company taking the other side of the bet. Here's how it works. An insurance company wants protection against a massive hurricane that will cause it losses above a certain level. It sells a load of catastrophe bonds worth, say, $500 million, which are bought by a bunch of specialist investment funds. The $500 million goes into a locked account and every year for the life of the bond – three years, perhaps – the investment funds receive an interest payment. At the end of the bond's term, the investors get their $500 million back. Unless, of course, the specified disaster happens. Then they lose a chunk of the money – or if it's a really huge catastrophe, the whole lot – and the insurer takes the cash to patch up its battered finances.

It might look at first glance like high-stakes gambling on humanitarian tragedy. But according to its proponents, this market could be a vital means of raising capital to offer widespread disaster protection on an increasingly dangerous planet. 'You see a lot of discussion about offshore tax havens, and this and that, but we have people here creating solutions for problems,' Greg Wojciechowski, CEO of the Bermuda Stock Exchange, tells me. 'In my career, there's not been many times where you see things with a really strong opportunity to do good for the world. And this asset class really, really does.'

On the third floor of an anonymous office block in central Hamilton I find the headquarters of Nephila Capital, the world's biggest ILS fund manager. Named after a silk-spinning Bermudian spider with a fabled ability to forecast hurricanes, Nephila manages nearly $10 billion of money for investors all over the world – a sum that it hopes will keep growing rapidly as awareness grows around this little understood corner of global finance. 'We're certainly not facing any limit,' says Barney Schauble, one of the firm's managing executives.

As a young Goldman Sachs banker, Barney was part of a team that created the world's first catastrophe bond in 1996 and worked to build the market over the following years. He moved over to the 'buy side' at Nephila in 2004, just in time for the industry's lift-off stage. After Hurricane Katrina devastated New Orleans in 2005, killing 1,800 people and leaving a record-breaking damage bill of $160 billion, catastrophe bond issuance surged as insurers struggled to meet a spike in demand for protection. Soon after that, there came another watershed moment in the sector's story: the 2008 global financial crisis. When the US subprime mortgage market collapsed, the contagion spread to almost every other market. Share indices nosedived all over the world, along with the prices of corporate bonds and all kinds of commercial and residential property. But the value of catastrophe bonds held firm. After all, a stock market collapse did nothing to increase the risk of a disastrous hurricane. The assets in which Nephila invested were, in industry parlance, 'uncorrelated' – a word that now sounded seductive to global pension fund managers who had just lost their bonuses thanks to the synchronised collapse of everything in their portfolios. And with the world's pension funds holding a total of $32 trillion, even a small shift in their allocation can move staggering sums.

Over the following decade, the amount of money in the ILS market grew five times over. About a tenth of that flowed to Nephila, where a crack team of meteorologists and statisticians was crunching numbers from new scientific research papers, to quantify climate change's impact on disaster risk. 'In other financial market sectors, a lot of the research is private,' Barney says. 'But if you look at climate change, there's a huge amount of people working for you all the time, every day, around the world, trying to make this information available. We'd be foolish not to take advantage of that.'

Even after losing money to a ferocious succession of storms and wildfires in 2017 and 2018, the ILS industry has continued to grow. Barney rejects the characterisation of it as a disaster casino. When his investments lose money, he points out, that cash goes to fund reconstruction in the wake of catastrophe. Moreover, this sector is bringing tens – and in the future, potentially hundreds – of billions in funding to provide badly needed disaster protection. That sounds like a possible answer to Ernst's fears of a world in which insurance becomes unaffordable for millions. Yet the effect on pricing can only go so far, Barney warns. The new capital will help provide insurance to people facing moderate risk. But for those living in places that are becoming downright hazardous, Barney reckons a cold blast of market logic is overdue.

Already, he points out, there are thousands of people in places like California and Florida whose homes are insured at subsidised rates by state-owned entities, because the cost of protection on the insurance market has become painfully high. 'At those prices, that state entity is almost guaranteed to lose money,' Barney says. 'Now, is that in society's best interest? Maybe we should spend our resources as a society on things like making sure our cities are defensible – and say to the person with the $20 million house in the woods in Napa Valley: "Look, either you pay an exorbitant amount every year for insurance, because that's what it costs – or maybe you should go somewhere else."'

It's the kind of politically toxic dilemma that will be forced on leaders by a rise in disaster risk that shows no sign of abating. As Barney and I speak in late 2020, California is battling the worst wildfires in its history, on the heels of an unprecedented outbreak of forest blazes in Australia. It has been a record-breaking year, too, for Atlantic hurricanes, with thirty major storms; previously, the records showed only one year with more than twenty. And even in November – usually the languid fag end of the storm season – the chaos is continuing. While I stroll along Hamilton's placid seafront, a hurricane of unusual force is pummelling a country with scant resources to counter it, financial or otherwise: the troubled Central American nation of Nicaragua.

* * *

Ciro and I set out early on a pair of Chinese dirt bikes from the hill town of San Rafael del Norte, throwing up dust on our snaking path through the clouds that settled overnight on the volcanic landscape. Ciro sets a breathless pace, slowing only to navigate an occasional ambling herd of cattle or knee-deep creek, as we pass between lofty slopes thick with plantations on our way to Miguel's farm.

These lush mountainsides have been sending coffee to the world

Created by thawing permafrost in northeastern Siberia, the Batagaika Megaslump has grown over the past few decades to become the biggest phenomenon of its kind in the world.

Sergey Zimov at Pleistocene Park, where the maverick scientist is working to save Siberia's permafrost by restoring ice age 'mammoth steppe'.

Tsho Rolpa, one of the biggest of the glacial lakes that have proliferated across the Himalaya as mountain glaciers shrink.

Rasmus Kristiansen hunting near Qaanaaq in northwest Greenland. The world's northernmost civilisation for centuries, his Inughuit people now find their hunting traditions threatened by declining ice coverage.

The great polar scientist Konrad 'Koni' Steffen. Four days after we met, he died on Greenland's ice sheet, having fallen into a crevasse created by warming temperatures.

Eko Atlantic, a vast new expanse of artificial land billed by its wealthy developers as a climate-secure enclave for prosperous residents of Lagos.

A street in Makoko, the water settlement on Lagos's lagoon sometimes called 'the Venice of Africa'. Locals say they are well placed to cope with rising sea levels and will simply need to use longer stilts for their houses.

Taro Island, the tiny capital of the Solomon Islands' Choiseul province, is shrinking as sea levels rise. The government is seeking funding to relocate the entire population.

Noel Traky, a young resident of Nuatambu in the Solomon Islands. His island has been cut in two by a rising ocean, dividing a once close-knit community.

Shahida Bibi, a ‘tiger widow’ in Gabura, southwestern Bangladesh. Having been ostracised in line with local tradition following her husband’s killing by a tiger, she is now struggling with the impacts of rising salinity levels.

Habibullah Mollah moved to the Dhaka slum of Bauniabadh after his shrimp farm in Gabura was destroyed by a cyclone. He now works as a rickshaw driver and hopes one day to return home and start another shrimp farm.

The Mose system separates Venice from the sea for the first time in history. After decades of planning, billions of euros of investment and one of Italy's worst corruption scandals, the storm barrier has been highly controversial.

Joanna Sustento protesting outside Shell's office in Manila. Her family members were among over 6,000 people killed by Typhoon Haiyan in 2013.

Thousands of Haiyan's victims were buried at this mass grave outside Tacloban, which has already fallen into disrepair.

Bermuda, the hub for the global catastrophe bond industry. This young asset class enables funds to bet billions on the outlook for natural disasters.

Fernando Cruz is one of many Nicaraguan coffee farmers hit by an unprecedented pair of severe hurricanes in November 2020.

since the nineteenth century, their fruit now found in trendy roasteries from Los Angeles to Taipei, where few patrons have much idea of the stories that lie behind their cup of Nicaraguan Arabica. Miguel's family have been growing coffee here for four generations – sustaining the business through decades of criminal dictatorship under the rapacious Somoza clan, then riding out the civil war that followed, when the forests of this northern mountain region were roamed by US-armed rebels against the new leftist Sandinista regime. Now they have been touched by another dark chapter in the nation's history: a pair of hurricanes unlike anything to hit Nicaragua before, forming a grand finale to the Atlantic Ocean's most active storm season on record.

As October 2020 drew to a close, it appeared Nicaragua had escaped the brunt of an exceptionally turbulent summer that had driven chaos across the Caribbean and the Gulf of Mexico, killing 165 people and destroying property worth nearly $40 billion. By that point in the year, the Atlantic storm season is usually petering out, with November typically bringing just a handful of minor events. But this was no ordinary November. As the month began, US meteorologists spotted a menacing low-pressure zone tracing a hooked path over Cuba towards Central America, gathering force with unusual speed. By the time Hurricane Eta slammed into the Nicaraguan coast, it had formed a spinning tempest with winds of 140 miles per hour. In 169 years of modern records, only twice before had a storm of such force formed in the Atlantic in November. The next one came precisely two weeks later. Hurricane Iota made landfall just 15 miles south of Eta's entry point, with winds of still greater force.

Between them the two hurricanes killed about 270 people and displaced hundreds of thousands, while destroying $9 billion of property from North Carolina to Colombia as they careered through the region. In Miguel's hilly northern region of Nicaragua, the hurricanes brought biblical downpours that caused havoc on the steep muddy slopes. A short distance to his southeast, 34-year-old Martha Lorena Hernández and her two small children were among a dozen people who died at the foot of a coffee plantation, after the saturated hillside morphed into a torrent of mud that buried them in their homes. Now, two months on from the storms, the area's farmers are midway through a harvest that, for many, will be far more meagre than they'd been banking on.

Ciro, to whose company Miguel owes about 5,000 US dollars, takes careful notes as Miguel, a tough-looking but diffident 29-year-old, describes the impact of the hurricanes on his land. The winds were like nothing people here had ever seen, ripping part of the sheet metal roof from his house. Far worse, he says, were the days of torrential rain that drenched his coffee groves, causing an outbreak of 'rooster eyes', a fungal infection that can devastate crop yields.

Miguel leads us through his plantation, its soil still a soggy quagmire, the plants packed so tightly that their branches form an unbroken mass on a slope as steep as a black ski run. Ciro traverses the terrain with practised ease; I fall on my backside a couple of times. Again and again Miguel stops to point out the tell-tale beige blotches tarnishing the leaves of his infected plants. He had been banking on collecting 9 tonnes of coffee beans in this harvest, but now the hurricane has destroyed 60 per cent of his crop. It'll be tough, he says, to make the next repayment on his loan.

'He was lying,' Ciro tells me later over chicken and beans in a roadside *comedor*. 'That wasn't a 60 per cent loss.' A native of the hill region, Ciro has spent the past two years cruising its rocky pathways on assignment for Financiera Fundeser, a Nicaraguan microfinance company that makes loans to small farmers. With thousands of clients scattered across Central America's largest and poorest country, Fundeser relies on Ciro and his three team-mates to spend their days plying Nicaragua's country roads, grilling clients and inspecting their farms to keep track of the state of the loan portfolio. And since the hurricanes hit, Ciro's work has taken on an urgent new dimension: helping Fundeser figure out how to deploy a windfall insurance payout.

For a big multinational insurance company, a small-time Nicaraguan farmer would be more or less the opposite of their ideal client for standard disaster coverage. The logistics of reaching his property to assess its value, and then any damage to it, would be prohibitive. You would probably outsource that work to a local partner – but in one of the world's twenty most corrupt nations, the risk of 'irregularities' in the loss estimates would raise eyebrows among executives sitting in distant New York or Zurich. And leaving all else aside, the farmer's meagre income means that the earnings from insuring him, even in the best possible scenario, would be negligible compared with far more lucrative opportunities insuring people and businesses in the rich world.

A few years ago, Fundeser's board realised that the almost total lack of disaster insurance among its clients was a major hazard. If a catastrophic storm hit Nicaragua, it could wipe out the harvest for many of the farmers who had borrowed from Fundeser, potentially causing mass defaults and a severe hit to the company's balance sheet. And the risk of such an event seemed to be rising to an unacceptable level. The Atlantic hurricane season of 2017 had been the costliest on record, with total damage surpassing $200 billion.

With the help of various intermediaries, Fundeser made a deal with XL, a Bermudian insurance outfit that was recently acquired by the French giant AXA. Fundeser's insurance policy with XL would have nothing to do with damage estimates. It would deal simply with the official weather data. This would be a 'parametric' policy – an innovation that's getting increasingly popular as global demand grows for climate-related hazard coverage. Fundeser would pay its annual premium, and if the rainfall in a given period fell outside the agreed range, then XL would pay out. (XL, meanwhile, has been getting its own protection by issuing catastrophe bonds.) November's disastrous downpours of course fell far outside the normal range, triggering a payout from XL to Fundeser of half a million dollars – cash that will enable this small institution to write off the season's loan repayments for the farmers worst affected by the storms.

Ciro's job is to find out who they are. The impact of the storms varied hugely, he's been discovering, from one valley to the next, with crop variety, soil type and local forest cover all helping to decide a farmer's fate. Word has got around of Fundeser's new approach, giving farmers an obvious incentive to claim huge losses – but, of the seven clients Ciro visits during my two days with him in the hills, Miguel is the only one who seems to have overplayed his hand. I can see why the others decided to stick to the facts. Ciro is a pretty imposing guy, with a rugby prop's build, a boxer's jaw and a natural scowl. Moreover, he knows what he's talking about and makes sure to show it, peppering the clients with detailed questions about their farming practices, dropping friendly tips on fertilising their fields and marketing their harvest. It saves time and trouble for everyone, he reasons, if they just talk straight.

When Ciro files his report to headquarters, he estimates Miguel's losses at 15 per cent – still substantial, but much less severe than those of others we meet. Old Don Fernando, with a cowboy hat and a white pencil moustache, grew his coffee in the traditional way, in a shady forest of towering tropical hardwoods. In that enclosed environment, humidity levels soared for weeks after the hurricanes, driving a fungal outbreak far worse than Miguel's. Then there's Uriel, who lost his entire onion crop, and Lionel, who watched $2,000 worth of tomatoes rot in his fields. Even for the venerable Don Fernando, Ciro makes sure to kick the tyres, inspecting every section of the plantation and discreetly gathering testimony from the neighbours. 'It's a bit like being a detective,' he grins.

Whether they realise it or not, those clients who pass his smell test are in line to benefit from a transformation that is sweeping one of the most powerful segments of the global financial sector. But for farmers on every continent, the dangers are mounting.

PART FOUR

Dry Land

CHAPTER 11

Climate change is a major challenge in wine production. Temperatures are increasing worldwide, and most regions are exposed to water deficits more frequently . . . Depending on the region and the amount of change, this may have positive or negative implications on wine quality. Adaptation strategies are needed . . .

– Cornelis van Leeuwen and Philippe Darriet, 'The Impact of Climate Change on Viticulture and Wine Quality', *Journal of Wine Economics* 11(1), 150–167 (2016)

MAULE, CHILE

A mile down the road, droves of cows and horses were escaping onto the highway, as the flames that had destroyed their fences began to eat a path towards Daniella Gillmore's winery. Daniella still bore the surname and milky complexion of the great-grandfather who had crossed the Atlantic a century before, but there lingered no trace of that Scottish heritage in her lilting Spanish, rapid-fire sentences peppered with the distinctive Chilean suffix *-po!* Now she stood guard in front of the whitewashed winery complex with its sloping terracotta roofing, and the vineyard that ran for a mile to its east, the vines packed in dense rows that maximised the potential for fruit production and, now, for rapid combustion. In the first weeks of 2017 – a year when adverse weather events across the wine map would bring the lowest global production since 1957 – Viña Gillmore was in danger of becoming an early casualty.

Here in the Maule Valley, 160 miles south of the capital Santiago, winemakers had long been able to count on healthy rainfall by Chilean standards. But now they were seven years into the worst extended drought on national record. In recent decades, the exuberant planting of pine trees by forestry companies had left Maule a mass of tinder for what would now prove the biggest wildfires in Chile's recorded history, ripping through over a million acres of the country's parched heartland.

For five days and nights the vigil continued as the flames danced near the edge of the estate, Daniella running shifts with her husband Andrés and two local friends. They felled trees and dug trenches on the property's perimeter, and filled with water the back-mounted packs normally used to spray weedkiller, primed for a desperate rearguard defence. At last, a faint scream rang out in the distance and built towards a deafening crescendo as a corpulent Ilyushin IL-76 aircraft approached, dispatched by the Russian government to help Chile tackle the emergency. The plane flew low overhead, 42 tonnes of water gushing from its rear onto the blaze. That, together with a change in wind direction, saved Viña Gillmore – but not the year's harvest.

The fire had failed to destroy the vines, but it left its mark on them. In the following weeks Daniella let her team proceed with the harvest as normal, pressing the grapes for tasting before fermentation. As she had expected, the samples all bore the acrid tang of the smoke and ash that had bathed the grapes for days. This year's vintage, she decided, would have to be destroyed.

* * *

The vines of Chile tell the modern history of the country – carried from Spain by the conquistadors, then tended by the Jesuit missionaries who followed them; bought up by the elite families who came to dominate the national economy, before being expropriated under Salvador Allende's abortive socialist regime; and finally emerging as a globally renowned symbol of Latin America's most impressive economic success story, with Chilean bottles now a fixture on wine lists across the globe. And as I discover on a journey along most of the length of this bean-shaped country – 2,650 miles long, with an average breadth less than a twentieth of that – Chile's vineyards give an insight into its future, and the future of agriculture in places around the world menaced by growing water problems.

I start in the country's north, in the driest region of the world outside Antarctica. Seven thousand years ago, the Chinchorro people began mummifying their dead in its northern reaches, where the moistureless sands preserved the bodies with stunning clarity until they started to rot in a modern museum. It drew General Augusto

Pinochet's secret police in the 1970s, seeking discreet locations to bury murdered dissidents, and later NASA researchers who deemed it the closest thing on earth to the Martian surface. And here, in the heart of the Atacama, a desert far older and drier than the Sahara, Wilfredo Cruz is making wine.

Off Chile's western shore squats a gargantuan high-pressure zone, creating a perpetual anticlockwise movement of air that courses in from the Pacific in southern Chile and then upward through the country, shedding water as it goes. The current is achingly dry by the time it reaches the Atacama, where it takes a left turn out to sea – pushing back any moist oceanic air that might try to drift inland. To the east, the Andes sealed the Atacama's desert fate, blocking humid flows from the Atlantic and the Amazon rainforest. Yet just enough moisture from beyond the mountains condensed and fell around the 5,400-metre ochre peak of Cerro Honar to support life for centuries in Wilfredo's village of Toconao – dribbling for 20 miles through a narrow creek to the settlement where farmers diverted it into small canals, growing grain and fruit in a region dismissed by outsiders as uninhabitable.

Then the outsiders became more interested, as this arid region became a hot territory in the global clean tech boom. Beneath the Salar de Atacama, a salt flat to Toconao's southwest, lies one of the world's great reserves of lithium, the key ingredient for electric car batteries. Mining group SQM is producing tens of thousands of tonnes of the mineral each year, pumping lithium-rich brine from under the salt crust into pools where the desert sun steams it into a dry white residue, destined for processing in China. Nearly all his old classmates have gone into mining, says Wilfredo, a barrel-chested 28-year-old. Farming in Toconao has become an old man's game, with few farmers under the age of 50.

Now, a new problem has emerged that threatens to finish off agriculture in the oasis, reducing it to a mere way station for mineworkers. With rainfall over Cerro Honar set to decline as climate change disrupts air currents around the Andes, Toconao's water supply is endangered. And there's little chance of topping it up by pumping water from underground aquifers. The extraction of brine by SQM's miners has been depleting the region's water table, according to communities to the south that rely on wells for drinking water.

But while the mining looks like a dire threat to people in other parts of the Atacama, Wilfredo is hoping SQM could prove an unlikely saviour of farming in Toconao. In an effort to woo support in the region, it's providing startup funding for a winemaking business, and has hired Wilfredo to run it, luring him home after years studying winemaking in Spain and at wineries near Santiago.

If agriculture is to survive in Toconao, Wilfredo says, standing in the shady fringe of one of the new vineyards, wine offers the best hope. Vines are much less thirsty than the quince trees that currently predominate here, he says. And if Toconao can carve out a niche in the international wine market – helped, presumably, by the novelty value of its extraordinary geography – the margins could be good enough to sustain livelihoods even at relatively low levels of production, as the declining water supply bites more gravely in the years ahead.

So Wilfredo is trying to marshal a motley gang of ageing farmers into high-end wine producers, several of them with minimal agricultural experience, like Don Manuel, a former truck driver with a roguish grin who has enthusiastically embraced Wilfredo's guidance. Some of the more seasoned farmers have been less pliable. They had been irrigating their crops in the way it had been done here for centuries – flooding the entire field and allowing the water to seep deep into the earth. Now Wilfredo expects them to use modern drip-irrigation techniques, with plastic tubes running the length of every row, spitting water through small holes next to each plant. It's a far more efficient use of the declining water resources, but must be carried out much more frequently than the old approach. Beyond the added workload, it's the sense of violated tradition that grates on the farmers, Wilfredo says. 'They're attached to doing things in their grandfathers' way,' he says. 'But if we don't change? No, this desert will die.'

* * *

Winemaking might look like a climate adaptation strategy in Toconao, but for Chile's more established wine regions the changing climate is throwing long-held practices and business models into question. The Maipo Valley, which surrounds Santiago, is the beating heart of the country's wine industry, having surfed the wave of export-driven Chilean growth that began in the 1980s. It's managed this through careful management of limited water resources: annual rainfall in Maipo is about half that of French regions like Bordeaux and is concentrated overwhelmingly in the winter, leaving soils at risk of drying out during the warm growing season. Still, with extensive irrigation using water from winter rainfall and Andean rivers, Maipo wineries have been able to turn out bottles of serious quality. They've spearheaded a boom in Chilean wine exports that are now worth nearly $2 billion a year, the fifth-highest figure in the world.

But when I arrive in the valley, its winemakers are reeling from years of miserably low rainfall that have dragged down production and cast a shadow over their future. 'You've come at a dramatic time,' says Felipe Tosso, as we look out over the baking floor of the huge reservoir that he's drained at unprecedented speed in an effort to offset this year's rain deficit, the worst of the decade-long drought.

Felipe is chief winemaker at Viña Ventisquero, which he has built, since it started production in 2000, into one of the biggest of Chile's new generation of globally popular wine brands. Felipe became aware of climate change's threat to the wine industry early in his career, as he watched the impact of heatwaves in Europe and wildfires in California. 'We saw that and thought – we are so lucky here,' he says. That luck soon ran out.

Since 2010, central Chile's winelands have suffered the worst extended drought in their history, with mean annual rainfall a bit more than half the average for the previous four decades. 'The so-called Mega Drought is the longest event on record and with few analogues in the last millennia,' some of the country's top climate scientists wrote in the *International Journal of Climatology*. As temperatures surge around the equator, the high-pressure zone that looms over Chile's climate is shifting south, scientists say. That means the life-giving northward air current that sweeps from the Pacific into southern Chile has to pass over more land, shedding moisture all the way, before it reaches Felipe's vineyards.

This past year brought just 82 millimetres of rain in the Santiago region, about a quarter of the historical average – indeed, lower than the average figure for famously arid Dubai. That's having a powerful impact on production volumes. At a dinner with other Maipo winemakers earlier in the week, Felipe found he was far from alone in expecting his harvest of Cabernet Sauvignon – the French grape that is the backbone of Maipo's wine industry – to be about half as big as the year before.

I walk with Felipe and Alejandro Galaz, his bearded deputy, along the floor of the winery, between oak barrels stacked six high, the air heavy with the tart aroma of fermenting fruit, towards the steel tanks holding the oozing purple mass that will become this year's vintage. We fill small glasses from a tap at the bottom of a tank. Alejandro swills a mouthful and spits it into a bucket from five paces with a sniper's precision. This year's vintage will need to be handled carefully, he says.

Quality as well as quantity can be threatened when the rains fail. Less water means smaller grapes, with sugar levels that rise rapidly as the season progresses. To avoid an excessively rich, alcoholic vintage, winemakers have to harvest the grapes sooner in the season – four weeks earlier than normal at Ventisquero this year. That leaves less time for other of the myriad characteristics of fine wine to develop – the complex mix of scents and tangs that leave wine writers grasping for comparators from blueberries to petroleum. It doesn't all have to be bad, Alejandro goes on. There's an argument that smaller grapes can yield a certain concentration of flavour. In any case, since it's now clear that new conditions are taking hold, winemakers have no choice but to adapt.

In a lab on one side of the winery, white-coated technicians are performing tests on just-picked grapes to assess their chemical balance. And Felipe and Alejandro are spending more time roaming the vineyards, tasting and inspecting the grape bunches to monitor the maturation process, which is far more rapid than in the past. They've started to see some strange behaviour from the stressed plants. 'This is the first time I've seen this kind of thing,' says Alejandro, holding a bunch of Cabernet Sauvignon grapes in a room above the lab. In a normal year in Maipo, a vine's grapes mature in synchrony, but here a surreal range of colour is present in a single bunch – some of the grapes plump and bursting in ostentatious purple, others lurking tight and green under Alejandro's bewildered inspection. 'What the fuck is *that*?' he mutters, mercilessly ticking off the deformities. 'What is *that* doing *there*? Look at *that*!'

* * *

Of the sixteen winemakers I visit during my journey through Chile, every one has reached the same conclusion: that the historical conditions that nurtured this industry have been replaced by a new, still evolving climate regime, which requires them to evolve in response – and they are doing so.

The managers of the Tabalí winery, 190 miles north of Santiago, have abandoned 70 of their 430 hectares of vineyards to focus their ebbing water supply on the most viable fields. All over Maipo, the big winemakers are exploring technological fixes – from humidity sensors in the fields to giant sponges that would reduce evaporation rates from their private reservoirs – and experimenting with different grape varieties that could be better suited for the increasingly harsh conditions. In Maule, Daniella – whose vineyards had yielded grapes for centuries with nothing but natural rainwater – is investing in irrigation, as well as creating firebreaks to guard against the next forest blaze.

Climate change is forcing deviations from tradition in the world's most venerable wine regions – notably France. Winemakers labelling their bottles with celebrated names like Bordeaux or Burgundy were always required to forego any kind of irrigation, which was considered a form of cheating, interfering with the wine's natural expression of its environment. But water worries are shaking up the stuffiest chambers of European wine. After a succession of droughts, France's INAO, the body that lays down the rules for those sought-after 'protected designations', has been gradually easing the irrigation rules. Traditionalists worried this was the thin end of the wedge, and were proved right. In 2018, the INAO said it planned to expand the small set of grape varieties permitted for use in the protected designations, to protect the industry's long-term future against climate change.

There's one coping strategy that's not an option for the prestigious French winemakers: to move. Their brand value is far too closely tied to the centuries-old fame of their home regions. 'You can't make Burgundy in Bulgaria,' smirks Marcelo Papa, chief winemaker at Concho y Toro, Chile's biggest wine producer and the fifth-biggest in the world. In contrast, Marcelo says, Chile's relative youth as a force in global wine, and the relative obscurity of its regions – few foreign drinkers know their Maipo from their Maule – gives it some welcome flexibility. Concha y Toro has built a scientific research centre 150 miles south of Santiago, where it's studying growth options in wetter southern regions beyond its Maipo homeland. At Ventisquero, Felipe and Alejandro are betting on robust production from their new coastal vineyards, which enjoy higher rainfall and cooler temperatures. And one company, with an eye to more extreme trends taking shape in the coming decades, is exploring opportunities at the very frontier of a warmer world – an initiative that brings me to Patagonia for the final stop on my Chilean trail, 1,600 miles south from Wilfredo's winery in Toconao.

The project is the brainchild of Miguel Torres, a slight, always impeccably dressed Spanish wine tycoon, who from his headquarters near Barcelona has become the industry's most prominent voice on climate change responses. Head of Spain's biggest wine company, he's been an influential figure in Chile since he started production there in the 1980s. Severe European heatwaves in the first decade of this century convinced Torres of the urgency of the challenge presented by climate change. He began reading up on the science and spending time with climate experts, gathering two hundred of them for a dinner in Catalonia ahead of the ill-fated Copenhagen summit in 2009. He set about greenifying his business's operations, installing solar and biomass energy hardware and a system to capture carbon dioxide from fermenting grapes. He hit the conference circuit, confronting audiences with primers in palaeoclimatology and the Arctic albedo effect. At least, he reflected, he'd be able to tell his grandchildren he did what he could to help protect their planet.

He's been no less focused on protecting his business, investing in irrigation at his vineyards and altering farming techniques to protect grapes from overheating. A team of his researchers is investigating old Spanish grape varieties, looking for strains long deemed substandard that could thrive in the changed climate. And he's been applying a similar logic to the geography of his holdings, as climate change reshapes the global wine map. Britain's winemakers, long seen as marginal eccentrics, have started winning prizes for their sparkling whites, as some of the top French champagne marques buy up land in southern England. Vineyards have started to proliferate in Sweden, Denmark, and Norway, including one near a shrinking glacier 150 miles northwest of Oslo. In Canada, winegrowers are venturing ever further north into Ontario and British Columbia.

As temperatures rise in Familia Torres's low-lying Spanish vineyards, it's been planting new ones high in the foothills of the Pyrenees, in locations that were always considered too cold for wine grape production but now look well placed for the future. In Chile, the company is moving south – it's bought 230 hectares of land in Itata, a region south of Maule that had been considered too rainy for high-quality production, and it's signed long-term purchase contracts with a cluster of new producers in Osorno, 550 miles south of Santiago, far beyond the latitude of Chile's traditional wine regions. 'Ten years ago you wouldn't have thought of producing wine in Osorno,' Eduardo Jordan, Torres's chief winemaker in Chile, tells me. 'It was all cows and sheep.'

The company is not just tracking where the band of territory suitable for wine production has reached so far. They want to know where it's heading next – a search that they've extended deep into Patagonia, that harsh territory of glaciers and extinct volcanoes that tapers into a cone at the southern extreme of South America, pointing towards Antarctica. Accounts of its desolate landscape have inspired writers since Shakespeare, and attracted fugitives from Butch Cassidy and the Sundance Kid to Nazi war criminals. Now it's arousing the curiosity of winemakers. Vineyards have started to appear on both sides of the Patagonian Andes, on land never before considered for viticulture. Miguel Torres's quest to understand the future of wine has propelled his team much to the south of rivals' new Patagonian operations, to Puerto Ingeniero Ibáñez – a mountain village level with the end point of New Zealand's South Island, far beyond the southern tips of Australia and South Africa.

I anticipate an underwhelming scene when I drive into the village, a cluster of low-lying buildings encircled by evergreens on the edge of Lago General Carrera, which forms a watery border with Argentina. The afternoon is clear and chilly, and beyond an idling ferry the lake's dark waters stretch towards the bare, jagged hillsides on the opposing shore.

I began the day with a visit to another of Familia Torres's experimental vineyards, 50 miles north of here, at a site called Los Cóndores, named after the swooping Andean scavenger now threatened by extinction. On a nearby hillside, orange-green needles were poking from the trunks of tiny saplings, the first section of a 5,000-hectare tree plantation to offset the company's carbon emissions. But within their small fenced-off area the vines were dead, brittle twigs inside green protective cylinders, having reached only a few inches in height before succumbing to the austral cold.

At Puerto Ibáñez, however, in a small field a few minutes' walk from the lake, shielded from the sharp mountain wind by a screen of pines, is a different scene: three rows of vines, about two dozen in each line, all bursting with pale green leaves eighteen months after planting. I approach with Armando Godoy, the former village head now supervising the project, and taste a few of the grapes hiding in crowded bunches amid the leaves – one row each of Riesling, Chardonnay and Pinot Noir, varieties bred in cooler parts of western Europe.

They're bitter enough to bring tears to my eyes, but the Torres team expects the flavour to soften as the plants mature over the next few years. And simply by yielding fruit they've beaten expectations. The company set up these experimental sites to monitor how quickly climate change could bring new regions onto the winemaking map. The appearance of these grapes suggests that's happening faster than expected.

Armando and I wander through the tiny vineyard as a black horse grazes in the adjacent field. His family moved to the region from Osorno nearly a century ago, when the Chilean government threw open the territory to settlers. There was no road connecting this place with the rest of the country, so they had to make a long loop through Argentina. Thereafter little changed for decades, with successive generations making a quiet living from raising sheep and growing apples and apricots. But now the growing seasons are warming and lengthening, with rain falling in place of snow, and the fruit harvests are swelling. 'Ask anyone here, they'll tell you the same,' Armando says. 'We're benefiting from this. Absolutely.' The birth of grape production here has underscored the rosy outlook for farming in this mountain enclave – even while winemakers elsewhere are forced to tear up their rulebooks and plough huge sums into adaptation measures. And as climate change continues to twist and distort the landscape of global agriculture, others face much bleaker prospects.

CHAPTER 12

Climate variability and extreme weather and climate events are among the key drivers of the recent increase in global hunger. After decades of decline, food insecurity and undernourishment are on the rise in almost all subregions of sub-Saharan Africa.

– World Meteorological Organisation,
State of the Climate in Africa 2019

AFAR, ETHIOPIA

Ali Mohamed woke beneath the long-thorned limbs of an old gerento tree, the remnants of his herd lounging nearby amid the black fist-sized hunks of porous volcanic rock that carpet this stretch of the Horn of Africa. After three years of withering drought that had killed all but his hardiest camels and goats, a recent burst of rainfall had given him a flutter of hope, injecting new vitality into dying beige clumps of grass. But that hope was shattered by the sight that greeted Ali when he opened his eyes that October morning: a dawn sky coated with a bulging mass of yellow locusts, gushing in their tens of thousands from behind the distant silhouetted hills.

A hundred miles to the south of Ali's home had once been that of Lucy, the upright-walking hominid whose discovery in 1974 reshaped the theory of human evolution. In the three million years since Lucy's brief life, this cradle of humanity has become one of our species' harshest habitats. In the north of Ali's Afar region lies Danakil, the hottest inhabited place on earth, a nightmarish landscape of pulsing lava pools and crusty mineral flats where hundreds of men mine salt in temperatures often approaching 50°C. Even in the more hospitable parts of this northeastern chunk of Ethiopia, only the most robust strains of vegetation grow in the flinty soil, watered by perhaps a single day of modest rainfall each month.

For centuries, the Afar pastoralist people have made this hostile terrain their home, amassing over countless generations a deep understanding of how to sustain life in a uniquely fragile ecosystem – where to drive the animals and when, to keep them alive without exhausting the vegetation that underpinned their long-term existence, while weathering the periodic attacks of cheetahs and hyenas. For shelter they built *ari* – dome-shaped homes of woven palm leaves stretched over a framework of tree branches. The houses' aerodynamic design meant they withstood the rough winds of the open Afar plains, and they could be quickly dismantled and loaded onto the backs of camels when it was time to move on. The camels and other animals gave meat and milk, and could be traded with outsiders for textiles and the long knives that Afar men wear on belts at their waists.

But the Afaris' ancient knowledge has failed to shield them from a shift to a new, still harsher climate regime. As a succession of increasingly powerful droughts ripped through their cattle population during Ali's youth, herders in his area abandoned that animal – once central to their culture – to focus on preserving their camels and goats, which had proved better able to survive extended spells of hunger and thirst.

Three years before my visit to Afar, Ali's father Mohamed died, leaving him 80 camels and 100 goats – enough to support the family he planned to build with his quiet, serious cousin Madina, whom he would marry a few weeks later. But before his death, Mohamed had spoken bleakly about the trends he saw taking shape in the region's environment. 'He said the climate is changing, and all these animals will end up dying,' Ali tells me now in the mounting morning heat as we crouch in the shade behind his *ari*.

Mohamed's forecast is now coming close to fulfilment. The year of his death brought the first of three successive failed rains, a calamity without precedent according to people in Kori, Ali's home district in central Afar. For a thousand days Ali and his teenage brother Moussa fought to keep their herd alive, roaming for weeks in search of increasingly scarce pasture, seeking updates on grazing conditions whenever they crossed paths with other herders on the same desperate mission. Afar herders had always been careful to pace their usage of the best grazing land to avoid exhausting it.

But now the ferocity of the drought left them with no option but to flock to any place where they had heard some grass still remained, before it, too, became barren.

Mabai, Ada-el, Adoalay, Abana, Galelo – Ali reels off the names of a string of areas that he traversed in search of pasture. In Musle, a plain half a day's walk from where we sit, the grass in Ali's childhood had been luxuriant enough to envelop an entire herd of goats, rising high above their heads. Now it was a parched, sullen dustbowl. Every few days, sometimes more frequently, one of the animals would silently lie down in the sun-cooked dirt, leaving Ali and the rest of the herd to walk on without it. Just three of Ali's eighty camels, and eighteen of the hundred goats, remained alive when, a few weeks before my arrival in Ethiopia, the rain finally fell.

And then came the locusts. All over the world, changes in temperature and rainfall are beginning to drive changes in the distribution of species – including insect pests that can pose a severe threat to food supplies. In 2018, an exceptionally intense pair of cyclones – linked, studies suggest, to rising sea surface temperatures in the Arabian Sea – drove heavy rainfall over war-torn Yemen, creating the damp conditions for its desert locust population to go into an orgiastic breeding frenzy. The next year, the wind from another strong cyclone swept huge swarms of locusts southward, across the 20-mile-wide Bab-el-Mandeb Strait that separates the Arabian Peninsula from Africa, into Somaliland and Djibouti on Ethiopia's northeastern edge. When the long-awaited rains fell in Ali's homeland the year after that, it was the locusts' signal to turn up in their millions.

They operated with military organisation, battalions of hundreds of thousands of insects peeling off from the main army to devour a single tree or patch of grass. They lingered for more than a week, each of them eating its own body weight every day, before moving north to virgin terrain. Behind them, in a final smirking insult, they left the residual scraps of vegetation littered with their droppings, khaki-shaded pellets that looked like food to the hungry livestock, which then began to fall sick with digestive problems.

As we speak, Ali's 18-month-old son toddles around us in an oversized orange T-shirt, on course to join the 41 per cent of Afari children who are physically stunted due to malnutrition. In his own childhood, Ali ate meat every day, sliced and hung in the sun to dry, along with chickpeas and lentils acquired from traders in the region. Now he, like everyone he knows, has been reduced to absolute dependence on government handouts. For weeks, he and his family have eaten nothing but *ga'ambo*, a flatbread that Madina makes from the wheat flour distributed by state workers who drive through the area in a truck every month. Behind him a scrawny camel calf waits restlessly for the return of its mother, gone for several days with young Moussa, whose turn it is to take the herd on a hunt for land less contaminated by locust shit.

A devout Muslim, Ali still speaks of restoring his herd to the size he inherited, if God wills it. Under the best conceivable conditions that will take a decade, he says. 'And if this kind of drought happens again – there's no way we can continue to live here.' But while the shifting rainfall pattern means a dire water shortage for Ali's people, it's bringing heavier downpours elsewhere in Ethiopia. And for one town at the other end of the country, that's become a crisis in itself.

* * *

Accompanied by his two camouflaged bodyguards, AK-47s swinging from their shoulders in case of an incursion by the Islamist fighters of Al-Shabaab, Said Ugas leads me through the ghostly streets of Mustahil, the town that he grew up in only to end up leading its abandonment. A short drive from the border with Somalia, the cheery, fast-talking 34-year-old guides me past the decaying traditional homes that once housed Mustahil's population, sloping grass roofs atop walls of bound branches sealed with animal dung.

Tracking back from the muddy brown river that curves around one end of the town, we pass the abandoned primary school, the empty hospital and government administration buildings, the collapsing mosque that flanks the once bustling marketplace.

Mustahil's origin legend centres on the outburst of a roving trader who passed through this slice of Ethiopia's far south when the town's founding fathers were building the settlement, more than two hundred years ago. Most people in what is now the Somali region of Ethiopia lived then, as many still do today, as nomadic herders much like the Afaris to the north – the only feasible way of life given the arid conditions in most of the region. But this group, led by one of Said's ancestors, had decided to build a permanent farming settlement, growing wheat and rice on the banks of the Shebelle River. The Shebelle burst its banks every year – providing fertile soil on the land that flanked it, but presenting a permanent hazard for anyone looking to build a town there. '*Mustahil!*' cried the merchant in his native Arabic. 'Impossible!'

Said's ancestors set out to prove him wrong and did so, learning how to build their houses and lay out their fields to withstand the floods that came with the rainy season in the Ahmar Mountains, where the Shebelle began 300 miles to the northwest. By the time Said's father took over as the town's twenty-third chief, the settlement had grown to a town of more than three thousand people.

And yet the merchant's jibe was beginning to look uncomfortably prescient, as Said's people suffered the flipside of the changes in East African rainfall patterns that had been driving the years of drought among Ali's community far to the north. While rain grew scarce on the stony plains of Afar, in the highland areas that fed the Shebelle it was becoming more concentrated in violent downpours, driving increasingly severe flooding in Mustahil. The townspeople had learned to live with shallow water entering their homes for a few days most years. But now the floods were coming up to five times a year, sometimes rising above head height, upending their lives for weeks at a time. Crocodiles revelled in the chaos, snatching animals and, on several occasions, people from the submerged streets and fields. Mohammad, the principal of the town's elementary school, learned to live with a grim new annual rhythm, as pupils were repeatedly swept away by the onrushing torrents.

As if this were not enough, Mustahil's people faced violence from multiple angles. On the other side of the nearby border with Somalia – with whose people they shared a culture, language and centuries of history – protracted drought had exacerbated already seething tensions that were tearing that nation apart. It's one of several countries where climate impacts already appear to be undermining political stability. Large numbers of young men in Somalia were unable to make a living from the parched land, and thousands were pulled into the ranks of Al-Shabaab, an Islamist terror group with links to Al-Qaeda. The group's violence soon spilled over the border. On a single day in 2013, Al-Shabaab terrorists roared into Mustahil and publicly beheaded six townspeople.

The threat from Al-Shabaab served as an additional excuse for the brutal tactics of Abdi Illay, the heavyset president of the region, who led a campaign of torture and terror against suspected separatists – including many from Mustahil. Abdi's inhuman, venal reputation meant that when he proposed a plan to relocate Mustahil to a new site away from the flooding danger zone, it failed to win the support of the suspicious population. But Said knew that on this point, if on few others, the tyrant was on the right track. Mustahil was becoming uninhabitable.

At last, in 2018, Abdi was swept from power. Said, now chairman of the town council, joined forces with his father to persuade the community that the time had come to move. About half the population endorsed the plan immediately and agreed to start moving out to the new site that Said had identified several miles from the river, on a wide, rocky plain scattered with thorn bushes. The remainder were still resistant. The new site would be too hot, they complained, with none of the shade offered by the trees that lined the streets of Mustahil. They would need to rely on wells for water and would have to travel much further to reach their fields.

With help from an engineer who had fled the fighting in Somalia, Said started planning the new town. The site was a sun-scorched desert, but it offered ample space for the settlement they laid out, with tin-roofed homes built on wide, unpaved streets. The other half of the population might fall in line once they saw the new town taking shape, Said reasoned.

In the end, the shift in sentiment happened far more quickly. As Said's team was starting to lay out the new town, one of the worst floods in memory hit, forcing almost the whole population to clear out immediately to the new site. A few holdouts lingered on in Mustahil – such as Adam, an intransigent old man who stayed on for days in his flooded home, and threatened to divorce his wife if she didn't do the same. He relented only after seeing a crocodile devour a donkey outside his front door.

Slowly the new town is coming to life, the merchants setting up their stalls in its centre, the mosque nearing completion, the ramshackle coffee shops hosting lively discussions among the townsmen in their long sarong-like *ma'awiis*. The move happened just in time, Said says. This year saw the most severe flooding on record – the river burst its banks five times – and while the move meant nobody was killed, the fields remain waterlogged, an entire year's harvest lost. Like Ali's family 600 miles to the north, Said's people must now rely on government aid until there's a long enough period without flooding for their fields to dry out. It's the first time in its history that this community has had to abandon farming for a whole year.

To go with its new location, the people have decided, their town needs a new name – one more upbeat than the original, whose black humour had lost its charm. Mustahil became Mustakim, an

Arabic word meaning 'straightforward'. The path ahead for this community – and for Ethiopia, with its dizzying mix of climate change impacts – seems quite the opposite.

* * *

From the Somali region, I was due to travel to Tigray, a drought-hit state in Ethiopia's far north. A day before my flight, travel into Tigray was suspended when the Ethiopian army launched a major assault on the state, accusing its leaders of plotting to break away from the country. There has since followed one of the worst humanitarian disasters in recent African history, with over a million people displaced and extensive reports of mass rape and massacres.

The bloody civil war has upended the stellar global reputation of Ethiopia's 41-year-old prime minister Abiy Ahmed, who had earlier won the Nobel Peace Prize for ending a long-running standoff with Eritrea – and who, like few other world leaders, has put environmental issues at the heart of his national agenda. A year after taking power in 2018, he led Ethiopians in a world record-busting day of tree planting, in which the country claimed to have planted 350 million saplings in twelve hours. Within four years, Abiy wants that number to reach twenty billion.

It's not just a public relations gimmick, energy and water minister Seleshi Bekele tells me at his office in Addis Ababa, Ethiopia's heaving capital. The tree planting, he says, is a key part of the strategy for tackling climate change, to which deforestation has helped make this country exceptionally vulnerable. A hundred years ago, as much as 40 per cent of Ethiopia was covered in trees. But as the population surged in the second half of the twentieth century, communities across the country felled huge tracts of forest for farmland, firewood and building materials, with tacit support from the government of the time. By the turn of this century, 90 per cent of Ethiopia's trees had been felled. The deforestation was a logical way for communities to boost their crop production in the short term, but it had painful consequences for long-term food security. Across the country, fertile topsoil was washed away from the denuded landscape in the rainy season – a factor behind the devastating famine in the mid-1980s, which prompted the Band Aid initiative by Bob Geldof's team of pop stars.

Now, global warming is adding to the danger. Across the Sahel, the arid belt of land that runs along the southern flank of the Sahara Desert and is already the world's poorest region, climate change has been dragging down average rainfall, stoking fears of creeping desertification. In 2007, an international plan was hatched to tackle this problem: the Great Green Wall, a vast reforestation scheme that would stretch from northern Ethiopia to Senegal on the other side of the continent, passing through central African states like Chad and Niger. But more than a decade in, little progress has been made in most of the member states – another example of the difficulty in marshalling multilateral action that has been seen in so many other theatres of the response to climate change. 'To be frank I don't know who is overseeing this initiative,' Seleshi tells me. 'Papers are written, talks are given. But the Great Green Wall has no champion.'

So Ethiopia has decided to move forward under its own steam. The trees will replenish sickly ecosystems across the country, Seleshi says, softening the impacts of both droughts and floods. And as well as protecting its farmers, he adds, the twenty billion trees will also sequester large quantities of carbon dioxide, taking Ethiopia closer to its goal of having a carbon-neutral economy. One of the great conundrums of the global climate change response surrounds the question of poor countries' development: the West got rich by trashing the environment, some argue, and is in no position to urge developing nations not to do the same. Ethiopia – one of the world's fastest-growing economies in recent years – wants to prove that green development is a real option, Seleshi says.

His nation has a green asset that sets it apart from most of its peers: the torrential flow of the Blue Nile, which rushes from Lake Tana in Ethiopia's north on a looping course towards Sudan and then Egypt. Ethiopia is already using the Nile to generate nine-tenths of its electricity from low-carbon hydroelectric plants. Admittedly, it's still a meagre amount by global standards. Ethiopia's total electricity capacity is 4.3 gigawatts – about 5 per cent of the figure for the UK, which has half its population. But that capacity is about to double thanks to a new dam taking shape near Ethiopia's western border, one of the biggest in the world. The Grand Ethiopian Renaissance Dam, or GERD, is set to jumpstart a new era of national growth, according to the government – but it also risks pushing international tensions in this already unstable region to new levels.

As analysts ponder the possible political consequences of climate change, one of the most alarming fields of study is the prospect of conflicts emerging between nations competing for increasingly scarce water resources. From the mountains of Tibet alone, there flows a web of rivers crossing multiple borders, offering rich scope for future tension between China, India, Pakistan, Bangladesh, and much of Southeast Asia.

But while those disputes are cause for long-term concern, the one between Egypt and Ethiopia over the Nile's waters has been reaching a new intensity. The two nations, which dwarf most of their neighbours, have had a simmering rivalry for centuries. One of the proudest chapters in Ethiopian history is the victory over Egypt in a three-year war in the 1870s, thwarting Egyptian ruler Isma'il Pasha's scheme to extend his control over the entire length of the Blue Nile. Isma'il Pasha's twenty-first-century successors in Cairo now have cause to regret his failure. With rising temperatures predicted to drive up its farmers' water needs, Egypt's government has grown alarmed about climate scientists' forecasts of increasingly volatile rainfall in the region that feeds the Blue Nile. In the event of a severe drought, they worry, Ethiopia could hoard water to ensure the smooth running of the GERD, with disastrous consequences for Egyptian farmers a thousand miles downstream.

While Seleshi leads negotiations with his Egyptian counterparts, tensions have been rising. Shortly before my visit, Ethiopia's government suffered a cyberattack from an Egyptian hacker cell called the Cyber_Horus Group, who replaced more than a dozen official websites with a picture of a grinning skeleton in pharaonic headgear, clutching a scythe and a scimitar. 'If the river's level drops, let all the Pharaoh's soldiers hurry and return only after the liberation of the Nile,' read a text beneath the image. Nationalist media pundits in Cairo have been beating the drum for an all-out military assault.

Seleshi smiles stiffly when I mention concerns about the possibility of war. 'They tried that in the 1870s,' he says. 'They lost.' The Egyptian complaints, he insists, amount to little more than opportunistic posturing aimed at wrangling concessions from Ethiopia and confusing the international community. But they seem to have made an impression on Donald Trump, now in the dying days of his presidency, who makes an extraordinary intervention two days later. 'Egypt is not going to be able to live that way,' Trump tells media at the White House. 'I say it loud and clear: they'll blow up that dam. And they have to do something.'

The remarks spark astonished outrage in Addis Ababa. On a shady terrace at Addis's Hilton Hotel, I hear a full-throated riposte from Zerihun Abebe Yigzaw, a diplomat who is one of Ethiopia's most high-profile negotiators on the dam, and who identifies so closely with his country that he talks about it largely in the first person singular.

Foreign governments 'might think that Ethiopia is a weak and poor country,' Zerihun says, 'that we'll be easily subdued and bullied. They are wrong. Who the hell are they to approve the filling of my dam?' In an age of increasing environmental dangers, the dam will be a launchpad for sustainable prosperity, he argues, protecting Ethiopia against a recurrence of disasters like the 1980s famine that colours the country's international image to this day. 'I don't want to be a humanitarian case for Bob Geldof,' Zerihun says. But as the impacts of climate change take hold, the challenges facing this country, and its already troubled region, look set to become still more formidable.

CHAPTER 13

Farmer income losses from climate change could be between 15 percent and 18 percent on average, rising to anywhere between 20 percent and 25 percent in unirrigated areas. These are stark findings, given the already low levels of incomes in agriculture in India.

– Indian Ministry of Finance, *Economic Survey 2017–2018*

MAHARASHTRA, INDIA

Even in the middle of January, a dense heat weighs on our chests as we shelter under a jagged acacia. Around us, dark hard-baked soil is scattered with the desiccated remnants of a failed harvest. It is quiet but for the incongruously cheerful chirrups of tiny birds hiding in the trees, and the occasional honking passage of distant motorbikes, or of yellow-roofed autorickshaws overflowing with passengers.

The farm is Tulsiram Kanere's. He thinks he's now about 63, and has worked this land since early childhood, when his father Vinayak would show him how to sow seeds, how to lead oxen to the fields and drive the plough. Tulsiram left school at 10 to help his father full time, and became the head of the family in his mid-twenties after Vinayak collapsed in the fields and died. Life has never been easy for farmers in Jalna district, in India's western state of Maharashtra, about two hundred miles inland from the giant metropolis of Mumbai. But the past few years, Tulsiram says, have been unprecedented in their harshness.

Indian farmers are among the world's most vulnerable to climate change – and nowhere more so than in Jalna, which has always been one of the country's most water-deprived areas. To its west, a hill range blocks moist air from the Arabian Sea. To the east, the land stretches for 600 miles. No major river runs anywhere near Jalna and, thanks in part to government corruption, irrigation is scant.

In recent years, stark changes in rainfall patterns have pushed Jalna's farmers into crisis. More often than not, the annual volume of rain during the monsoon season, from June to September, has been falling far short of the historical average. And what rain does come, comes increasingly through sudden huge downpours that do more harm than good, the water saturating the earth and running off, taking fertile topsoil with it. As conditions worsen for farmers in Maharashtra, an epidemic of suicide among them has attracted national concern. Before the year is out, another 3,927 will have taken their own lives.

Under a soft white cap, Tulsiram's face is thickly creased, with furrows bursting like butterfly wings from the edges of his eyes, the product of a ready grin and years of sunburn. A tasselled lilac scarf hangs over a short-sleeved white tunic; below that a coarse white sarong, hitched up on one side to reveal a shin zigzagged with scars from sharp crop stalks. In the September harvest, Tulsiram says, he gathered less than 75 kilograms of millet – from land that had produced more than 2 tonnes in good years. The yields of his maize and cotton crops were almost as bad. 'It was said that there would be good rain, so I spent a lot on seeds, fertiliser, everything,' says Tulsiram, his grin effaced, gnarled hands kneading at his waist.

He waited in vain for the rain that would bring a profitable harvest, gazing at clouds that skulked overhead for days on end but never burst. Then, one day in August, the weather forecast said there would be little rain for the rest of the season. 'That,' he says, 'was when I realised everything I had invested was lost.'

Tulsiram is not at the bottom of India's wealth ladder. With 10 acres of farmland, he's much better off than the millions of landless poor. But the succession of droughts has been steadily eroding his resources, forcing him to start selling his animals and rack up growing amounts of debt.

His travails are mirrored by increasing numbers of farmers worldwide – a struggle of fearsome importance. The world's population is set to reach nearly ten billion by 2050. With growing affluence in middle-income countries, global food demand will rise even more quickly. Yet as climate change takes hold, even traditional levels of output are becoming impossible to achieve in many places. Around the world, crops are being pummelled by repeated droughts and temperatures well outside the normal range. A few months after I meet Tulsiram, a new US study will reach a shocking conclusion: climate change has already begun reducing the yields of major crops, snatching thirty-five trillion calories a year from the global food supply.

Now Tulsiram, and millions of other Indian farmers, are at the centre of one of the most urgent problems of our age. Will humanity manage to feed its surging population from land swept by rising temperature, and all that comes with it? And are we willing to tolerate radical, even unsettling, technological interventions to do so?

* * *

'India is the first of the hungry nations to stand at the brink of famine and disaster,' William and Paul Paddock wrote, more than half a century ago. 'By 1975 civil disorder, anarchy . . . and chaotic unrest will be the order of the day.' The brothers' bestselling book *Famine 1975!* was one of several high-profile prophecies that swept the intellectual circuit in the 1960s, proclaiming a wave of starvation closing down on India and other poor countries struggling to feed their burgeoning populations.

Tulsiram's Jalna district played a part in helping India to avoid this fate. Even as the Paddocks issued their dire warning, a local engine oil salesman named B. R. Barwale was starting to make his mark in the agricultural sector. Barwale's Mahyco was the first company set up to tackle India's critical food shortage using new techniques discovered in the US. Scientists had crossbred grain types to produce 'hybrids' with dramatically increased yields, particularly when boosted by chemical fertilisers.

This 'green revolution' took India by storm and saved it from mass hunger – or at least from perpetual dependence on foreign food aid

– by enabling huge increases in the amount of food that could be produced from a single field. But now, climate change is beginning to undo that progress. On current trends, the Indian government has warned, national farm output will fall by 15 to 18 per cent this century – in a country currently adding nearly fifteen million more mouths to feed each year.

'The only reason I'm optimistic,' Barwale's daughter Usha Zehr tells me, 'is because technology is available.' She's guiding me through Mahyco's hulking cream headquarters, set amid well-watered lawns seven miles from Tulsiram's fields. At the entrance, orange marigolds are draped over a portrait of Barwale, bald and smiling benevolently, who died eighteen months earlier. Usha has inherited her father's cheery aspect, grinning and chuckling behind her rimless glasses. But at this corporate campus surrounded by struggling farmers, she is deploying some of the most controversial techniques of modern biotechnology.

In one brightly lit room, filled with the hum of expensive machinery, Usha shows me the elaborate torture inflicted on small leaf fragments to extract their DNA for analysis. In the final step, the battered tissue is spun in a centrifuge sixty times a second to reduce it to its constituent parts, including what the scientists are after: a vial of pure deoxyribonucleic acid, the stuff of life.

Usha was born in 1964, the same year Mahyco was incorporated. Barwale used his wealth to give her the higher education he never had: she left India at 17 to start fifteen years at US universities, and married Brent Zehr, a fellow plant scientist. The couple moved to Jalna in 1996 to take charge of Mahyco's research department, where they led the company's forays into the emerging science of genetic manipulation. With growing precision, scientists were now able to take a gene from one species and insert it into another. Want to tweak your new corn strain with a gene from another plant, a fish, even a human? It can be done.

After Brent's death in 2007 from a rare cancer, Usha continued to drive Mahyco's work in this field. And as climate change's impact on Indian farmers grows increasingly severe, she's become convinced that genetic engineering will prove crucial, creating new super-crops able to withstand increasingly extreme conditions.

For some experiments, Usha's team deploys 'gene guns', which use compressed air to fire foreign DNA into a dish of plant tissue. For others, they use bacteria – inserting an alien gene into a simple microbe that then infects plant cells, transferring the gene in the process. The latest addition to the arsenal is the CRISPR gene editing system, which uses chemical reactions to slice open DNA and insert invented sequences never found in nature.

If all this sounds to you like dangerous God-playing, you are not alone. Usha and her entire industry are under fire from an energetic global campaign, fighting to block genetic tampering that activists say could present unforeseen dangers to human health and entire ecosystems. But global biotech researchers are defying the controversy, pressing ahead with work to apply genetic engineering against climate threats. Mahyco is among the groups working on GM cereals that could survive extended drought, and strains that can cope with the salty soil brought by rising sea levels in places like Gabura. Others are focusing on crops equipped to resist soaring temperatures, and the upswell of pests that they will bring. Mahyco's partner Bayer, a global biotech giant, is already marketing Droughtgard, a genetically modified maize designed to survive intense water stress.

At Mahyco's labs, we walk through a room pulsing with intense lighting, packed with tall shelves bearing countless petri dishes of genetically engineered crops. Energised by hormones, the plant cells grow initially into calluses, clumps of tissue no prettier than those formed on feet by ill-fitting shoes, which then sprout shoots and leaves for the inspection of technicians. It's hard to tell which of the newborn blobs of matter is composed of rice cells, which of wheat or banana. But these ugly little life forms, Usha believes, could end up spawning crops with the traits needed to survive unprecedented climatic upheaval – and protecting the flow of food to millions of the world's poorest people.

'People say, God didn't mean for that to happen,' she says with a throaty laugh. 'I say, God gave me a brain.'

* * *

To its opponents, the biotech industry is not a source of salvation for struggling farmers but a key driver of their problems. And the concerns about agricultural distress in eastern Maharashtra have made Usha's and Tulsiram's region a focal point for these activists – chief among them Vandana Shiva. A sexagenarian with a grey ponytail and a beatific smile, Vandana has become known as the 'rock star' of the global fight against Big Biotech, delivering thunderous speeches to adoring activists from Hamburg to San Francisco. Her fans include Prince Charles, who has installed a bust of her at his Gloucestershire estate.

For Vandana, the entire green revolution was a grievous mistake, locking farmers into a system of expensive industrial agriculture that saddled them with debt while degrading their soils. Now, she believes, that grim saga is culminating with climate change: the outcome of a pathological economic system that has modern agriculture – with its heavy machinery and fossil fuel-derived fertilisers – at its heart.

As one of the drivers of global warming, Vandana argues, industrial farming – especially in its most advanced form of GM agriculture – can form no part of the solution. She urges farmers to turn instead to traditional seeds and organic techniques honed over centuries, which she promises will protect crop yields from climate stresses far better than the products of the biotech industry.

'We already have the seeds of climate resilience,' she tells me. 'And if Gandhi could pull down the British cotton empire with a spinning wheel, then we just need a simple seed to save us from this empire over life.'

We are sitting in the New Delhi offices of her activist group Navdanya, housed in a few flats on a residential street in Hauz Khas, an upscale area that grew around a medieval water tank and is now best known for its trendy restaurants. From outside come the grunts of passing vehicles and the yells of hawkers, and the acrid air of the world's most polluted megacity.

Like Usha, Vandana was raised in a prosperous family and studied in North America – but there the stories diverge. On vacation from her doctoral course in philosophy at the University of Western Ontario, in her home district in the foothills of the Himalaya, Vandana became involved with local women who were trying to stop deforestation by hugging trees en masse. Since then she's become an icon in the movement against industrial degradation of the environment, and a missionary for traditional farming practices. But it's for her fierce opposition to genetic engineering that Vandana has become most celebrated, with her warnings of massive health risks and a future of 'corporate dictatorship' over food.

Her defiant position has put her on the other side of the debate from some of the biggest players in the environmental movement – notably Bill Gates, whose huge philanthropic foundation is pouring billions into helping farmers adjust to climate change, and who has repeatedly called GM technology a vital weapon in the fight.

'Bill Gates is a big mischief maker,' Vandana tells me sharply. But that is precisely how her many critics see her: as a firebrand without relevant expertise (her PhD thesis was on the philosophy of quantum theory), who has slowed the implementation of a potentially vital technology. To observers like the environmental writer Mark Lynas, she has become a 'demagogue', guilty of the same offences committed by opaquely funded climate change denialists: making grand allegations unsupported by scientific evidence.

Leading bodies such as the World Health Organisation have found no evidence that GM foods currently on the market are any less safe than those created through traditional breeding processes. A handful of studies over the past twenty years, claiming to have found proof that a GM-heavy diet can harm animals, have been roundly dismissed by leading scientists as methodologically flawed.

Vandana, however, believes those scientists have avoided seriously investigating the safety of GM food, scared off by the power of 'Big Ag' conglomerates. 'The philosophy is: don't look, don't see, and you can declare it safe,' she says. In private conversation she still deploys her full range of rhetorical devices, her pace of speech alternately racing and crawling, often following a point with a momentous pause, staring at me to gauge the effect of her words. And the effect, over the years, has been profound.

Vandana's impact has been most obvious in India, where she has taken part in several government consultations on biotech regulation, with top politicians writing glowing forewords for some of her many books. No GM food crop has ever been approved in India – only a pest-resistant strain of cotton – and, for all their long-term hopes, industry figures like Usha see little chance that the government will change its wary stance in the near future.

But Vandana's influence reaches far beyond her homeland. She has helped to rally opposition to GM technology around the world – most notably in Europe, where there is still no large-scale production of GM crops. The European situation has had major ripple effects in Africa, where governments have resisted the technology, reasoning that concerns about contamination could hit vital exports to Europe. Scientists in countries like Tanzania and Kenya, battling to soften climate change's impact on food production, have railed against the restrictions. If they are right, anti-GM activism is crippling one of the world's best hopes for avoiding mass hunger in regions like Sub-Saharan Africa and South Asia, where many are already chronically underfed.

Vandana is unrepentant as she leans forward onto the table that separates us, clutching a steel beaker of water. To my right is a painting of a bucolic Indian rural scene being despoiled and plundered by a huge hand, its besuited sleeve bearing the names of big industrial groups. 'I'm witnessing as I travel around the country how severely we are on the threshold of massive disaster,' Vandana says, her voice almost a whisper. The climate crisis is only strengthening her resolution against the companies she accuses of using it as cover to tighten their tyranny over farmers. Saving Indian agriculture from global warming, she is certain, means steering away from the siren call of biotechnology, and undoing the life's work of Usha's family. It's a struggle set to run throughout this century, with growing bitterness and urgency as the world's temperature and human population rise together in a precarious dance. Its twists will reverberate far beyond the biotech research labs and boardrooms, beyond the activist conferences and rallies, to the fields of farmers on every continent, and the mouths of billions who depend on what they produce.

PART FIVE

Meat

CHAPTER 14

Northern midlatitudes, over central Eurasia in particular, have experienced frequent severe winters in recent decades. A remote influence of Arctic sea-ice loss has been suggested . . . Our results strongly suggest that anthropogenic forcing has significantly amplified the probability of severe winter occurrence in central Eurasia . . .

– Masato Mori et al., 'A reconciled estimate of the influence of Arctic sea-ice loss on recent Eurasian cooling.' *Nature Climate Change* 9, 123–129 (January 2019)

UVS, MONGOLIA

Tuvtsengel rose before dawn, as always on this first day of the new year, which wandered on the modern calendar according to calculations already long established in the age of Genghis Khan. Leaving his wife Pamaa to slumber on their narrow bed, he roused his son Nyamkhuu from the floor next to the extinguished iron stove at the centre of their *ger*, a circular wooden framework covered with thick felt that protected them against the deadly Mongolian winter cold, dozens of degrees below freezing.

The air slapped their faces with familiar violence as they stooped to pass through the doorway on the *ger*'s southern edge. Each clad in a thick, fur-lined winter *deel* with the hem brushing the tops of his cowhide boots, they trudged through thick snow hardened by the cold to the texture of sandpaper, towards Bayankherkhen, the sacred mountain, where a handful of other herders had gathered. The congregation made six clockwise revolutions of the stone cairn that crowned the hilltop, before burning an offering of juniper plants. They could see for countless miles over the jagged landscape, just starting to be freed from the new moon's blackness by the purple glow of the incipient sunrise. Their voices rang out over the hills, pleas to the mountain spirits to end the curse of a deadly winter.

The calamity had begun more than six months earlier, during a summer of drought in this remote section of Uvs, Mongolia's northwesternmost province, where 80,000 mostly nomadic people are scattered across a barren, mountainous territory more than twice the size of Belgium. At their summer camp near Lake Uvs, a two-day journey to the north of their winter base, Tuvtsengel's goats and sheep fed far more thinly than usual on the parched grass, and when his clan set to scything hay near the lake's shore, Tuvtsengel's share amounted to only two loads of a rented pickup truck. As he made the journey back to the winter camp, he knew this haul was dangerously thin.

Since he and Pamaa married thirty years earlier, they had grown their herd from a few dozen animals to 500 – halfway to the 1,000 tally that would put them among the country's elite herders. In previous winters, he had sent some animals to relatives in Dörvöljin, 190 miles to the south, where conditions tended to be milder. Using one of the long-wave phones that had proliferated among herders a few years before, he called that family; they agreed to help but said they could take only a fifth of Tuvtsengel's herd. They were already swamped with requests from other relatives in Uvs, similarly nervous about predictions of an unusually harsh season.

With their remaining 400 animals, the family hunkered down for the winter, hoping that the forecasts would prove wrong. But one day in late December, they were engulfed by dense fog that seemed to sweep in simultaneously from Lake Uvs to their north and Lake Khyargas to their south. With the fog came snow, tumbling towards them in angry blizzards, draping a heavy blanket over their camp and the hills around it, which were already sparsely vegetated after the summer drought. Tuvtsengel continued taking the animals out to the hillsides, driving them with yells and whistles towards stretches where they might eat, but as the snow continued to amass, the flock found it increasingly hard to penetrate it for even a few mouthfuls of grass. Brief warm, clear swings merely added to the crisis, as the sun melted a thin layer on the snow's surface that swiftly turned to glassy ice when temperatures dropped again.

Tuvtsengel tried to fend off the animals' starvation with his slender hay stockpile, while keeping enough in reserve for the remaining winter months. When January brought a sudden plunge in temperatures, a level of cold exceptional even to herders and animals accustomed to 30° below zero, the deaths began in earnest. At the end of one day in the frozen pasture, he returned home cradling a sheep that had lain down in the snow in surrender to its fate. He placed it with the others in the timber night shelter, knowing full well he would find it dead in the morning.

At the new year ceremony, two Buddhist monks arrived to add their voices to the herders', chanting Tibetan prayers against animal death. But days later the winter entered its bitterest phase yet, with still more violent snowstorms, just as Tuvtsengel's ewes entered the birthing season. His hay store now empty, he watched helplessly as life faded from the next generation of his herd, unable to gain nourishment from the milk of their ailing mothers. He added their tiny bodies to the stacks of bony corpses downhill from his home. He gave one an autopsy, hoping to gain a better understanding of the catastrophe wrecking his livelihood. In its final hours, the starving lamb had filled its stomach with dirt.

* * *

Purevdorj's army boots crunch in the snow as we walk towards the long steel-roofed hangars that hold vital relief for herders ahead of what's forecast to be a harsh second half of the winter. Inside them are stacked thousands of bales of hay and mountains of grain fodder in white sacks trucked over the nearby Russian border. This is one of five emergency relief stores that Purevdorj manages in and around Uvs's provincial capital of Ulaangom, where treacherously icy streets link the orange portico of its communist-era theatre with a rash of seedy karaoke bars and, clustered on its outskirts, the *gers* of former herders partway through the process of urbanisation.

It's -16°, cold enough to sting exposed hands within seconds, and Purevdorj is dressed accordingly, in thick camouflage fatigues and a black fur hat with earflaps. But by the standards of mid-winter Ulaangom, this is oddly warm, Purevdorj says. The predictable seasons of his childhood have given way to strange volatility in the Mongolian climate, with each autumn now bringing grim foreboding of the phenomenon that has ravaged herds across the country: *dzud*.

In any era of Mongolian history it would have been a lucky herder who reached old age without experiencing a dzud – a lethally harsh winter with large numbers of animals killed by unusually cold temperatures, exceptionally heavy snowfall, or a combination of both. But over the past two decades the frequency and severity of dzud events has been increasing – with disastrous consequences for many nomadic herders like Tuvtsengel.

At the start of this century, three consecutive years of severe dzud across huge areas of the country killed nearly a third of Mongolia's animals. Another national disaster followed in the winter of 2009–10, killing a fifth of the country's livestock in a single year. In Uvs almost every winter now brings a dzud alert in some part of the province, Purevdorj says, after leading me into the warmth of his office adjoining the hay reserve and making a pot of tea.

The pattern is consistent with scientists' warnings about the knock-on effects of melting Arctic ice, of which the volume lost each year is enough to fill 120 million Olympic swimming pools. This leads to more moisture in the air above the ocean, which can then be dumped as snow as it moves south. Winter snowfall in Mongolia has increased 40 per cent since 1960. Just as dangerous is a second effect: the ice melt has been distorting long-established circulation patterns, weakening westerly winds and enabling more air from the Arctic to reach Mongolia and linger there. It's a powerful combination, increasing the chances of heavy snowfall at some points in the winter, and spells of extraordinary cold at others. Similar factors may be driving an increase in severe winter storms in Europe and the United States, including the one that threw Texas into chaos in February 2021. For Mongolia, the problem is compounded by conditions in the rest of the year: with the country's average temperature up more than 2° since 1940, increased evaporation in summer is contributing to a worsening trend of droughts during the crucial growing season.

For the past four years Purevdorj has headed the Emergency Management Agency in Uvs, responsible for administering rations to save the province's animals from mass starvation in times of dzud. His long face is topped by thinning hair and his grave, thoughtful expression creases to reveal gold lower incisors when he smiles – something that happens rarely as he describes the frustrations of his work. 'The system has gone all wrong,' Purevdorj says.

Uvs' size and complex geography – wide open plains flanked by hulking volcanic mountain ranges, its two large lakes overlaid in winter with ice thick enough to drive a truck across, the Siberian steppe to its north and the sands of Xinjiang to the south – means that winter conditions can vary wildly between the 19 *soums* into which the province is divided, and even between the 93 *baghs* that make up the soums.

Some of the worst dzud impacts, Purevdorj says, have been in soums that were known for their relatively mild winters. Sagil soum, to the west of Lake Uvs, had long been spared the deep snows and lethally cold temperatures found elsewhere in the province, meaning many of its herders had not seen the need to build the protective shelters that graced winter camps elsewhere. That proved disastrous in early 2018, when the soum was swept by snow in volumes that took the herders by surprise, followed by a plunge in temperatures well below -40°. Animals died in droves, and Purevdorj's team were powerless to save them, the waist-deep snow preventing their trucks from reaching the stricken camps.

As the dzud season approaches again, with heavy snow forecast for the weeks ahead, Purevdorj is openly exasperated with a situation where annual crisis has become the norm. All too often, by the time his trucks arrive bearing relief, the animals it's intended to save are already dead, or too weakened to survive. As the winter conditions become more dangerous, he says, more must be done to build resilience – through stockpiling of fodder at the local level, and construction of covered shelters – instead of relying on emergency handouts.

Purevdorj's diagnosis has a second, more subversive strand. Over centuries, Mongolian herders learned to live within the limits of their punishing environment – growing their herds only to levels that the flinty soil would support, rotating between pastures to avoid exhausting the fragile tracts of grass. The ancient systems were exploded in 1924 when Mongolian revolutionaries declared the nation a socialist republic, under the effective control of Stalin's Soviet Union. The herders were forced to organise into collectives, accountable to the planning ministry in Ulaanbaatar. Just as in Soviet factories, maximising output became the organising principle, with herders given production targets set out in five-year plans.

When the communist regime was replaced by a modern market economy, it turbo-charged the push for production. The meat processing industry became controlled by private companies that grew fat on transporting flesh to feed the hundreds of thousands of people living in Ulaanbaatar, as well as to cater to rising meat demand in China's now fast-growing economy. Instead of aiming for subsistence like their ancestors, herders now saw a large flock as the only path to financial prosperity, enabling them to buy vehicles and send their children to university. For those less inspired by material gain, the government continued to honour the most productive herders with an official 'champion' title. 'Every herder wants to be a champion,' Purevdorj says. 'And all the kids then want to emulate their fathers. It becomes a disastrous situation.'

The result has been a dangerous loss of stability, Purevdorj says. The number of livestock animals in the country has surged to seventy million, triple the number in the communist era. That means herders are now pushing towards the limit of what the land can support, even in a good year – and in bad years, which are now ever more severe and frequent, they're flirting with disaster.

Purevdorj gives another of his mirthless smiles when I ask what could address the problem. An obvious solution would be a punitive tax on herds above a certain size, he says – electoral suicide in a country where a quarter of people raise livestock, and where big hitters in the meat industry wield meaningful economic clout. 'They can't eliminate this system,' Purevdorj says, 'because the herders are the voters.'

It's a troubling line of reasoning: democracy and capitalism have raised living standards across Mongolia, giving rural families access to improved healthcare and goods from mobile phones to motorbikes. Yet with those gains has come a set of incentives that undermine the sustainable way of life at the core of Mongolian culture – something that's now being cruelly exposed as climate change impacts take hold. In his own way, Purevdorj embodies the paradox, struggling to fulfil his appointed duties in fending off threats to a stock of animals that, he is convinced, has bloated out of control. But for others in the region, the upheaval is throwing up new opportunities.

* * *

Tegshjargel and her husband Tservee have lived for nearly a decade in their long single-storey house, but it retains some of the trappings of their old ger – the stove chimney rising through its centre, the brightly painted wooden chests against its northern wall bearing photographs of deceased forebears. Still, the gaudy chandelier hanging from the living-room ceiling, the widescreen Samsung TV playing dubbed Hollywood films, and the Hyundai pickup truck parked outside are testament to the upward mobility of a duo who, through a mix of luck and instinct, have ducked and weaved their way through the dangers of dzud more dextrously than any other family I meet during my travels in Mongolia.

She high-cheek-boned with a regal posture, he with curly hair and twinkling blue eyes, the couple grin at each other as they recall their teenage romance. The children of herders in Sagil soum, northwest of Ulaangom, the couple met while boarding at the school in the village that serves as the administrative centre for Sagil's few hundred herding families. They married in 1995 and started out as herders in their own right, with an initial flock of animals gifted by their parents.

In the aftermath of the 2000 dzud, they started to focus increasingly on trading. Reading the signs of approaching peril, they had stockpiled more hay than most other herders and were left in a far stronger financial position than peers who had lost swathes of their flocks. In the years that followed they expanded from wool into meat trading, finally giving up their own herd and moving to Sagil's village to focus on the more lucrative business. Plenty of rival traders came in from Ulaangom and beyond, often with deeper pockets, but Tegshjargel and Tservee were known as a dependable local option. And in early 2018, when Purevdorj ruled out any chance of getting the state-aid trucks to Sagil gers engulfed by snow, it was to Tegshjargel and Tservee that the herders turned.

The calls started coming in early March, from herders caught out by the force of the blizzards attacking their ailing livestock. Tegshjargel and Tservee took to the blue pickup, making multiple runs each day to the ger camps hiding in sheltered dips in the hillscape, where the herders greeted them with desperate gratitude. They kept going for two weeks, shovelling themselves out of snowdrifts on their way to the anxiously waiting families, carrying goats and sheep back to the yard behind their house, where the snow turned red as they slaughtered them, dozens each day, cutting their throats before skinning them, cutting out the bones and stuffing the meat into plastic sacks.

They arrived at the Ulaangom sausage factories to receive offers of just 900 tögrögs per kilo of meat – 33 US cents, about a third of the normal rate, with prices depressed by a surge of supply from stricken herders. The trader couple decided to forego any profit from the exercise, passing on the takings to the herders in their entirety. As well as neighbourly sympathy, hard business logic was involved. Having proved the value of their partnership in the worst of times, Tegshjargel and Tservee now had a formidable edge over their deeper pocketed rival traders.

In the aftermath of the crisis, Sagil's herders decided to organise themselves. They're in the process of forming a co-operative – a middle way between the restrictive socialist collective structure and the open-market system that succeeded it. The idea is to soften the impact of fluctuating conditions on individual herders, as they band together to buy supplies and to find markets for their meat. For the latter, they'll be relying on Tegshjargel and Tservee, who are set to become the exclusive sellers of meat from the Sagil co-operative. 'These terrible conditions have made people realise they must work together,' Tegshjargel says as I demolish a bowl of her *buuz*, boiled lamb dumplings glistening with oil. Like all the other big upheavals in their career, she and Tservee appear to have played it to perfection – showing the kind of nimble adaptability that will be vital if Mongolia's age-old herding culture is to survive into a new era.

* * *

I wake in my sleeping bag on the floor of Tuvtsengel's ger, a strip of deep purple sky just visible through a gap in the centre of the roof, as he loads palm-sized dung patties into the stove just starting to gnaw at the morning chill.

It's been a year since the family's winter crisis, which left them fearing the devastation of their herd. Of the 400 animals at their camp that winter, half died – their bodies piled in rigid towers to await the spring thaw, when Tuvtsengel doused them with petrol and incinerated them. But the 200 that survived, together with the others that found refuge with the relatives to the east, were enough to keep the family clear of the 100-animal mark viewed among herders as the poverty line.

Tuvtsengel pulls on his dark green deel, fastening it with a piece of fabric wound round the midriff in the style worn centuries ago by the feared Mongol horsemen who built one of the world's great empires, stretching from Korea to Hungary. Outside, in the strengthening morning light, he takes a pitchfork and tosses out hay for the small cluster of cows milling on the slope above his home.

In the months leading up to this winter, at their lakeside camp the family gathered twice as much hay as the previous year. But for most of the summer, until the rains finally came in the last days of July, it had looked like another year of drought. With their depleted resources, the family would have been hard-pressed to buy enough fodder to sustain their herd – especially given the hefty tuition fees of their daughter Nyamsuren, training to be a surgical nurse in Ulaanbaatar.

Her brother Nyamkhuu, too, had eyed a life away from the herd. At 17 he was conscripted to serve two years' military service, the bulk of which he spent manning a guard post at the Russian border. Tall and broad-shouldered, Nyamkhuu got a taste for military life and considered becoming a career soldier until his parents told him to forget the idea. Now 24, Nyamkhuu is reconciled to the herding life and determined to sustain the culture of his ancestors. 'I'm proud to be living on this land,' he tells me. 'We will overcome this.' But since last year's events, his parents have started to wonder whether he will be able to do so. Many more years like that, Pamaa says one day, stitching a cowhide boot by the stove, and herding here will become impossible.

The Mongolian new year is ten days away, and, while conditions have been manageable so far this winter, heavy snow is forecast for the coming weeks. The real danger season, Purevdorj told me in Ulaangom, is just beginning. But today the air is crisp and dry as Tuvtsengel leads his herd out to pasture, driving them with whistles and shouts and waves of his wooden staff, down into a snow-caked valley and over a black stone ridge, to the flinty hillside that will be their pasture for the day. It looks a rough prospect for a hungry sheep, tiny yellow shoots lurking under a cape of icy snow, but they steadily work their way across it, using their hooves to scrape away the snow from one small patch at a time.

From a peak, Tuvtsengel watches them, against a backdrop of the barren hills rolling towards the frozen surface of Lake Khergil – a landscape intimately familiar since his boyhood, now a theatre for a fast-evolving swirl of deadly hazards. Yet while the shifting climate is creating new hazards for animal farmers like Tuvtsengel, the meat industry itself is one of the biggest drivers – as I am to see for myself, in temperatures 50° warmer, on the other side of the world.

CHAPTER 15

It is a fallacy to say that the Amazon is the heritage of humanity and a mistake to say – as the scientists attest – that the Amazon, our forest, is the lung of the world.

– Jair Bolsonaro, president of Brazil, at the United Nations General Assembly, 24 September 2019

URU-EU-WAU-WAU INDIGENOUS TERRITORY, BRAZIL

If the invaders had left Awapy any room for doubt about the earnestness of their death threats, it vanished the night they stabbed Ari in the neck.

Born in the years after government officials pulled the Uru-Eu-Wau-Wau into sustained contact with the wider world in the 1980s, Awapy and Ari grew up in a community ravaged by modern Brazilian violence and diseases, which had quickly wiped out two-thirds of the tribe's 250 people. They learned Portuguese and received a modern education – while still holding fast to the old traditions, not least when each married the other's sister in the symmetrical union required by tribal custom.

Now, in the steamy southwest of Brazil's Amazon region, the brothers-in-law had found themselves in the front line of the battle over the world's biggest rainforest – confronted, it seemed to them, not only by criminals and vagabonds, but by one of the country's mightiest industries, and by the new holder of the highest office in the land.

On a warm, sticky morning towards the end of the Amazonian dry season, I follow Awapy along a roughly cleared trail of hacked-at undergrowth. Around us rise pale-trunked mahogany trees and hefty kapoks from whose bases protrude huge vertical fins that famously offer hiding places for prowling jaguars. A relentless string of piercing wolf whistles tracks every stage of our progress – the unmistakable greeting of the screaming piha, the world's second-loudest bird and one of tens of thousands of species living in this rainforest, the most biodiverse place on earth.

'You can always tell when a path has been made by the whites,' Awapy says. 'They make such a damn mess.'

Demarcated by a government decree, the Uru-Eu-Wau-Wau Indigenous Territory comprises 1.9 million hectares of lush jungle, surrounded on all sides by evidence of the most exuberant campaign of forest destruction that humanity has ever achieved. When the Brazilian military government pushed itself into Awapy's people's lives, it was part of a broader strategy for the Amazon – with a priority of making the region contribute more to the national economy. And the key to that agenda, it had decided, was beef.

After the construction of the 2,500-mile Trans-Amazonian Highway in the 1970s, aspiring cattle farmers headed west in their tens of thousands, drawn by government financial support and the promise of legal rights to any land they cleared if they could show they had farmed it for a full year. The world's biggest rainforest began to vanish at an unprecedented pace. Since 1975, over 300,000 square miles of it have been lost – an area larger than Turkey – with most of this land turned into cattle pasture. That may have been a disaster for the environment, but the generals' vision for agribusiness was realised handsomely. By 2003, Brazil had overtaken Australia to become the biggest beef exporter in the world.

A grave-faced 28-year-old, Awapy is slightly built and gently spoken, but he has inherited a certain warrior spirit from his father Arimá, who killed an official from FUNAI – the government agency for indigenous affairs – with an arrow during the fraught days of early contact. At 15, Awapy led a dozen older Uru-Eu-Wau-Wau to confront a father and son who had come from a nearby town to fell trees on the reserve. They detained the younger man, grudgingly releasing him after his father returned with a large group of men, one of whom jabbed a loaded shotgun at Awapy's chest. In the following years, the pressure on the reserve eased. That reflected a broader slowdown of deforestation in the Amazon, as the left-wing government of Luiz Inácio Lula da Silva responded to a growing domestic and international outcry by ramping up enforcement actions against illegal deforestation.

But by 2018, the signs of resurgent danger had become too strong for Awapy to ignore. Invasions of the reserve had been picking up in line with an increase in illegal clearance in the Amazon as a whole, amid the political turmoil that had engulfed the country after a historic corruption scandal. And now the front-runner in the presidential election race, instead of promising to reverse the trend of illegal forest clearance, was stoking the ambition of its perpetrators.

An army captain in the days of military rule, Jair Bolsonaro went on to cut a marginal, eccentric figure on the back benches of Brazil's parliament for nearly three decades. His biggest claim to fame was his grotesque bullying of a female lawmaker whom he called a 'slut' and 'not worth raping'. Incredibly, Bolsonaro then ran for president on a family values platform – and was propelled to victory on a wave of fury at a mainstream political class oozing with corruption scandals.

A key support base for Bolsonaro was Brazil's agricultural barons, whom he wooed by promising to galvanise farming growth – not least in the Amazon. In a move that sent shivers through the region's tribal groups, he vowed that decisions on indigenous lands would no longer be taken by the specialists of FUNAI, but by the agriculture ministry. Bolsonaro openly stated his belief that the country's indigenous reserves were 'an obstacle to agribusiness' and needed to be whittled away. 'The Brazilian cavalry was really incompetent,' he had once told Parliament. 'The competent ones were the American cavalry, who decimated their Indians, and now they don't have that problem in their country.'

One day Awapy received a WhatsApp message from a white acquaintance in a nearby town. 'When Bolsonaro takes over,' it said, 'you guys are fucked.'

Ten days after Bolsonaro's inauguration on New Year's Day 2019, Awapy received word of the biggest invasion on the reserve for years. On the territory's eastern edge, dozens of men from nearby towns had entered and started felling the forest with chainsaws. Awapy arrived to find the head of the nearest village in a resigned mood after conversations with the loggers, who had bragged of their implicit support from the new president. 'There's nothing to be done, Awapy – the land is lost,' the older man said.

Awapy had other ideas. With his smartphone, he captured images of the invasion that he passed on with GPS coordinates to the police, forcing them to take action. Seemingly tipped off, all but one of the gang scarpered in time to avoid arrest, leaving behind them a wreckage of fallen virgin forest and bullet holes in the rusty sign that marked the entrance to the tribe's land.

Still, the episode opened Awapy's eyes to new ways of protecting his people and their land. With support from several non-profit organisations, he and his brother-in-law Ari started building the younger generation of their tribe into a battalion using technology to defend the Amazon. Awapy recruited his 15-year-old niece Borep to handle the semi-professional camera the group had acquired, to document the illegal activities on Uru-Eu-Wau-Wau land for authorities and the wider world. Puré, an energetic 23-year-old, was put in charge of the GPS tracker to record precise coordinates for the sites they discovered. The youngsters also drafted Jurip, a cousin of Awapy's in his forties, who was well versed in the more traditional forms of jungle work, claiming to be able to smell where the whites had been. The team got their most powerful tool in the form of a lightweight drone made by China's DJI. Now they could scan large areas of their territory from 800 metres above the ground, and quickly identify the sites to move in on.

That on-ground action was still Awapy's favourite part of the work. A few months before my visit, he and Ari led a raid on a camp deep in the jungle, where outsiders had been preparing to fell a swathe of forest. 'They'd built a wooden hut and shelters,' he recalls. 'There were mattresses, coffee and rice, axes, chainsaw chains, machetes, a bunch of bottles full of shotgun shells.' Having removed the ammunition, the team burned the entire site. 'I have to say I enjoyed it,' Awapy grins.

But the young task force was making some bitter enemies. For months, contacts outside the reserve had been passing on warnings to Ari and Awapy that they would be killed if they continued the resistance. A few weeks after their fiery raid on the squatters' camp, Ari left the reserve on his motorbike for a meeting in the town of Jaru. Ambushed, the schoolteacher and father of two was left to bleed to death in a roadside ditch.

With his first child on the way, Awapy considered quitting his leadership role and settling down to a quiet life tending his crops of manioc and banana. But he concluded that was not an option. 'If we don't resist,' he tells me, 'they'll destroy the forest – and us, with it.'

A couple of months before my arrival, Brazilian television aired footage taken by Awapy's team of their arrest of a trespasser on their land. The man appears terrified of his captors, with their warpaint and longbows, who treat him with firm civility until the police arrive. As they lead the defence effort, the younger generation is assuming wider authority in the tribe, Awapy says. 'The older people are relying on us – they don't speak Portuguese or know how to deal with the authorities. They ask us: "Why are the whites invading our land? We don't invade theirs."'

Ari's was not the last violent death amid the struggle for this territory. A few days before I arrived in Brazil, I read international news reports about the killing of Rieli Franciscato, the regional head for FUNAI. Walking deep in the reserve, Franciscato ran into members of one of the tiny tribes that, unlike the Uru-Eu-Wau-Wau, continue to live in isolation from the rest of the world. One loosed an arrow that struck him above the heart.

This was yet another blow to Awapy, who was scheduled to see Franciscato the very next day – a long-awaited engagement where he planned to seek more support from the agency for his team's operations. Relations with FUNAI had improved since his father's killing of the agent a generation earlier; Franciscato was respected for his long service in the area and commitment to the indigenous cause. But Awapy refuses to blame the uncontacted tribe. 'The invaders have been in their area and have probably been attacking them,' he says. 'They must have thought Rieli was going to do the same.'

Around us, the forest floor is shrouded in the heavy shade cast by the jungle hardwoods, their arching branches jostling for space in the canopy high above. But a short distance ahead, the vegetation becomes abruptly shorter, paler, infested by grasses alien to the rainforest. A drone flight reveals the scale of the damage. A rectangle several miles long is being carved in stages, its creators burning the forest and scattering grass seeds in the ashes.

The project may take years to complete but there's no question as to the vision behind it. There's no forest left to clear outside this northern patch of the reserve, where the cattle pasture runs right up to the border. Demand for Brazilian beef is soaring as customers in Europe and North America are joined by the growing middle classes in China and other developing economies. In the year before my visit, Brazil exported a new world record of 1.8 million tonnes of beef – enough to make twenty billion Big Macs. For some local farmers hungry for expansion, Awapy's territory is the obvious option.

'People need to think carefully about how much Brazilian beef they eat,' Awapy says. 'This industry is out of control.'

* * *

Hips cocked and arms crossed over a torn red T-shirt, Ezio stands at the heart of his 50 hectares of rainforest, between a patch reduced to smouldering detritus and a thriving stretch of lush vegetation awaiting the same fate. His one good eye shines with good-humoured defiance as he defends his small role in the Amazon's destruction. 'People like me slash and burn because there's no other way to get by,' he says. 'You need to find a way to survive. And what do we have to do that here? I'll tell you: cattle.'

It's been another busy fire season in Ezio's district of Boca do Acre in the southern underbelly of Amazonas – the biggest of Brazil's twenty-six states and home to the bulk of its surviving rainforest. Up in the country's northwestern corner, bordering Peru and Colombia, Amazonas escaped the worst of the early years of deforestation as farmers worked their way in from the east. In satellite images, the rainforest cover of the states to its lower right – not least Rondônia, home to Awapy's people – has been reduced to a clumpy mess like the ribcage of a mangy dog, but Amazonas still presents an almost flawless cloak of basil green. Look more closely, however, and you'll see a string of pale smears spreading like inkblots around the towns near the state's southern border. As the cattle industry runs out of room to expand in the states bordering Amazonas, the rainforest's last bastion is undergoing an accelerating wave of illegal forest clearance, in what Brazilian media are calling *o novo arco de desmatamento* – the new arc of deforestation.

Ezio's town of Boca do Acre was built overlooking the point where the Acre River flows into the Purus, which winds a tortuous 550 miles to the northeast before finally melding with the Amazon proper. Narrow wooden boats clutter the near banks, waiting to be steered away on trading journeys; a slope rises sharply to a promenade where black-winged turkey vultures crowd patiently around the weaker street cats. With its streets of clapboard dwellings, Boca do Acre has what might be called a tropical Wild West feel. As I dine at a pavement table one evening, a man staggers past bleeding from a stab wound to the midriff, followed by a wailing woman who clambers into the boot of a car that screeches away to take him to a doctor. But it's in the forest surrounding the town that the real action is taking place.

Like many settlements in the southern Amazon, Boca do Acre spends much of the dry season bathed in tangy smoke that floats over from funereal plumes rising in the distance. One Sunday I follow one of those billowing towers – turning off the southbound highway onto a rutted dirt road flanked by enormous pastures belonging to a local cattle tycoon known as Tãozinho. The road ends at an unpainted wooden house, where a footpath runs into an infernal wreckage of ashes, soot and scattered bursts of flame, amid which I run into Ezio, machete in one hand, jerrycan in the other.

Land like this can be traded legally but clearance of rainforest is tightly restricted, on paper – and to burn trees on this scale is a serious offence. Not that Ezio seems troubled by legal niceties. With a loan from his new father-in-law, he's taking the plunge into cattle farming after years on the fringes of the industry – first doing odd jobs on ranches, more recently working for a business that sells them equipment. 'I've seen how those guys just grow money in their fields,' he says. 'All those big farmers were an inspiration to me. I remember one of them telling me this whole area would become a huge cattle-field. Now I look around and see his vision becoming a reality.'

The destruction of the Amazon rainforest has enabled thousands of Brazilian cattle ranchers to make their fortunes, and may yet do so for Ezio, too – but it's a grave and growing problem for the planet. The Amazon is one of the world's most potent carbon sinks, removing more than 400 million tonnes of carbon dioxide from the atmosphere each year – enough to cancel out the annual emissions of the UK or Australia. But every dry season, this absorptive capacity shrinks as huge areas of the forest are burned, a process that in itself sends hundreds of millions of tonnes of carbon into the sky. And soon, according to a growing body of research, that relentless shrinkage may take on a life of its own.

* * *

Before heading to the Amazon, I drove from São Paulo to the satellite city of São José dos Campos to meet Carlos Nobre, Brazil's best-known climate scientist. Just as with the Siberian permafrost and the sea ice disappearing off Greenland, Carlos believes, we are pushing the Amazon towards a spiralling feedback loop, in which the loss of one piece of forest triggers the disappearance of another. We are getting dangerously close, he's been warning, to a 'tipping point' where the forest will begin to destroy itself, without any need of assistance from the likes of Ezio.

Since the 1970s, scientists have understood that vast rainforests like the Amazon don't simply emerge in areas that happen to receive a lot of rain. Instead, by driving increased rainfall as they grew, they created the conditions for their own flourishing over millions of years. In today's Amazon, each molecule of water is recycled over and over again – captured by a tree's roots, escaping through transpiration from the microscopic stomata in its leaves, before falling in a raindrop on another part of the forest.

Tall and rangy with a grey walrus moustache, Carlos has spent much of his career exploring how this life-giving dynamic could be affected by the ongoing destruction of the rainforest. Using the powerful atmospheric circulation models that have revolutionised climate science, he and his collaborators have made some alarming findings. His latest estimate: when the impact of human deforestation is combined with that of global warming, and of the dryer forest's increased vulnerability to wildfires, the tipping point is likely to be reached when 20 to 25 per cent of the original rainforest has been cleared. From then onwards, Carlos predicts, the Amazon will shrink under its own momentum, with two-thirds of it becoming a degraded savannah within less than fifty years. When we meet at his home in a quiet gated estate, the figure is above 17 per cent – and climbing daily.

'Ten years ago I would not have thought we'd already be approaching the tipping point of "savannisation",' he says. But the scientific data has continued to pile up: a lengthening dry season across much of the Amazon; declining rates of carbon absorption; rising mortality among the tree species most vulnerable to dry conditions. 'It's no longer possible to say this is natural variability.'

Yet instead of heeding the scientists and taking urgent measures to protect the Amazon, Carlos says, the Bolsonaro government seems to be leaning in the opposite direction. 'You've got organised crime behind this deforestation,' he says. 'And because of the political speech of the current government, they feel the risk of being caught and punished is reduced – which is true. They are feeling empowered.'

During my journey through the southern Amazon, Bolsonaro makes a video appearance at the United Nations General Assembly, haranguing the body for a second straight year about the injustice of the claims against his administration. 'We are victims of one of the most brutal disinformation campaigns,' he says, accusing international institutions of backing 'shady interests . . . with the objective of harming the government and Brazil itself.'

But I find plenty in the region to reinforce fears about an accelerating slide into lawlessness. Those behind the deforestation seem to have little fear of punishment – Ezio, for example, is convinced that the environmental agency Ibama, under pressure since Bolsonaro slashed its budget, won't bother pursuing a small-time offender like him.

The next day I meet Valdení Porto Miranda, known to his friends as Polaco ('Polack') for his fair complexion, who runs a cattle farm on 436 hectares of illegally cleared land, and has received fines from Ibama totalling nearly US$300,000. Sitting outside his small wooden house with his wife and three children, the born-again evangelical Christian says he hasn't paid a penny of the penalties and is betting that the overloaded agency will forget about him. 'By the grace of God, I have faith they'll send my file to the archives,' he says.

Polaco's faith might be rewarded, judging by my conversation with a disheartened Ibama agent at a roadside diner in western Rondônia. His group casts an imposing presence, escorted by heavily armed police in camouflage fatigues, but his unit is now just a quarter of its former size after Bolsonaro's budget cuts. 'Of course our morale is suffering,' he mutters under the chatter surrounding us.

A couple of hundred miles further east, a platoon of armed police blocks a newly built dirt road running through a protected forest reserve. Behind him, says the gangly sergeant in charge, his comrades are in a standoff with a 150-strong gang who have been felling a chunk of the forest. One of his unit has already been shot in the stomach. 'These cowards have brought their wives and kids with them,' the sergeant says. 'They've been here for weeks. You see, they've got protection from local politicians with farming interests.'

The police managed to catch one of the squatters, who taunted them with their powerlessness to have him punished. 'He said, you can arrest me as many times as you want – I'll be back. And he's back now, sure enough.'

The strangest tale of all comes from the badlands of Boca do Acre – that of Sebastião Gardingo, the big-time farmer known locally as Tãozinho, one of whose pastures I passed on the way to Ezio's fire. During my stay in the district, Tãozinho is under siege by prosecutors who accuse him of using local policemen as a private militia, to defend land he was illegally clearing for pasture.

The policemen – one of whom went by the nickname 'Death' – were paid by Tãozinho in cattle as well as cash to protect the land against impoverished locals who wanted to claim it for themselves, according to the indictment. One of this latter group was almost killed by shotgun fire after approaching the area being cleared by Tãozinho's bulldozers, the prosecutor said. As the man lay in hospital with wounds to his head and chest, his mother got a call from the leader of Tãozinho's police militia, warning that his men would continue to provide this 'service' to the local tycoon.

Early on my last morning in Boca do Acre, I pay a visit to

Tãozinho's ranch, passing under the magnificently horned bull's skull that hangs in his driveway, aiming to catch him before he heads out to supervise his workers in the pasture. While the policemen accused of working for him remain in custody, Tãozinho has been bailed from prison due to his age – though his 73 years sit lightly on him when he appears and spends the next hour giving his side of the story in the shade of an outbuilding.

'That road was like a mud pond when I came here from Minas Gerais – there was no power, nothing,' says the pioneer of the area's cattle industry, who now claims to provide a livelihood for over a hundred families. And while all the pasture here was originally jungle, Tãozhino insists, farmers like him no longer see the need to destroy forest to expand – instead focusing on boosting profits through scientific breeding techniques. He's been made a scapegoat, he protests, for the real culprits – the gangs of desperate, impoverished squatters in cahoots with corrupt local officials. 'There's nothing I can do – these guys are violent, shooting each other all the time,' he says. 'You have to understand, there's a war on here.'

While he contests the militia-running charges, Tãozinho's business is thriving, dispatching thousands of cattle a year. I picture the final consumers of the animals, perhaps in Beijing or Barcelona, oblivious to the colourful origin story of their prime grass-fed beef. I ask Tãozinho where the meat from his herd ends up going.

'I sell to everyone, all the processing companies,' he says. 'But where it goes after that – I have no idea.'

* * *

Imagine a factory where the latest techniques of hyper-efficient twenty-first-century manufacturing are deployed to build not cars or electronics, but cattle. Hundreds of workers at ceaselessly rolling assembly lines, each slotting into place a specific bit of bone and muscle. At last, pale off-white skins are wrapped around the completed animals, which wander out into concrete corrals before boarding a fleet of lorries that will carry them to farms in the surrounding region.

Play that mental movie backwards and you have an approximation of the Diamantino factory of JBS, a giant of the Brazilian economy. Every day at this plant in Mato Grosso state, south of Amazonas, 1,500 cattle are slaughtered and processed – disassembled, really – with an unblinking focus on slick efficiency. Hundreds of workers stand in front of conveyer belts, each responsible for carving a different cut of beef, armed with a hook in one hand and a knife in the other, all sporting chain-mail aprons to prevent them from accidentally stabbing themselves in the midriff. All around them, cleaning staff – dressed, prudently, entirely in red – sweep frantically. Almost nothing from a dead animal is allowed to go to waste. Their bones are ground down and used for fertiliser, the desiccated blood sold as fish food, the fat captured to make soap and biodiesel. Even the semi-digested contents of their stomachs are put to use, collected and burned to generate the building's heat and electricity.

This plant, along with more than thirty others scattered across Brazil, is at the heart of one of the world's great food empires. When 19-year-old José Batista Sobrinho set up his cattle slaughter business in 1953, at his family farm in the central state of Goiás, he was able to deal with just a handful of cattle each day. But as Brazilian beef production surged from the 1970s onwards – chiefly from deforested parts of the Amazon – his eponymous company rose with it, opening a string of factory-scale operations and building deep ties with major food retailers in the US, Europe and Asia.

JBS is now by far the world's biggest beef producer, with annual revenue of $51 billion: double that of its customer McDonald's, and second only to Nestlé among global food producers. But the past few years have been tumultuous for JBS. In 2017, the founder's sons Joesley and Wesley – until then JBS chairman and chief executive respectively – were both imprisoned after their holding company admitted paying bribes worth $150 million to 1,800 Brazilian politicians to secure cheap funding from state entities. Under new management – although the disgraced brothers remain the dominant shareholders – JBS's business has continued to flourish, booking record profits.

Yet as if the upheaval at the top were not enough, JBS's managers have been grappling with a fiendishly complex problem that, with global investors now keenly interested in their portfolio companies' environmental records, can no longer be sidelined: the alleged processing, in JBS factories, of cattle reared on illegally deforested Amazon land. 'Climate-changing deforestation and raging fires are clearing the way for more JBS cattle even now,' Washington DC-based campaign group Mighty Earth says in the month of my visit to the company. 'No responsible enterprise should be doing business with JBS.' The tide of criticism presents a mortal threat to JBS's global reputation, at a time when investors and consumers are increasingly repelled by unethical sourcing. It is Márcio Nappo's job to rescue it.

We meet at JBS's headquarters in downtown São Paulo, as rain falls in sheets from an ashen morning sky onto the hulking building's glass façade. A fast-talking trained economist with a dry sense of humour, Márcio had spent fifteen years working in Brazil's booming soy industry when he was approached by JBS in 2012. The company had just been humiliated by the release of a Greenpeace report accusing it of driving Amazon deforestation, timed to grab the attention of media covering a major environmental conference in Rio de Janeiro. Clients expressed alarm – especially in the UK, where JBS supplies the own-brand beef products of supermarket chains like Sainsbury's. The task of allaying their suspicions would fall to Márcio, hired as JBS's first director of sustainability. Investment in technology, he immediately realised, would be the key.

Much like any other major commodity, the market in Brazilian cattle is driven by intensely competitive daily trading. Every day, JBS cattle buyers purchase nearly 35,000 animals from a pool of about 100,000 suppliers, about half of whom are in the Amazon region. To meet its commitment to avoid procuring cattle from illegally deforested land, Márcio knew, JBS would need to make sure it was keeping track – on a daily basis – of the tree cover on the land of every one of those 50,000 Amazon farmers. Otherwise there was always a chance that one of its buyers would make a deal with a farmer the day after he burned a patch of rainforest, before the clearance had been detected by the company.

This was possible thanks to INPE, the Brazilian space research institute, which had made freely available high-resolution satellite imaging of the national territory. Márcio commissioned Agrotools, a São Paulo tech startup, to build an artificial intelligence system using INPE data to detect deforestation, cross-checking changes in the colour of the landscape with shifts in temperature. All JBS's suppliers in the Amazon region had their territory marked on the map, and any found to have cleared rainforest illegally would be blocked automatically. So far more than 11,000 have been struck off, more than a tenth of the supplier base. 'We're monitoring fifty thousand farms on a daily basis, in an area bigger than Germany,' Márcio says. 'It's the biggest supply chain monitoring system in the world, and probably the most sophisticated.'

And yet JBS continues to come under fire as the destruction of the Amazon accelerates. In an ominous sign soon before my arrival in Brazil, Nordea – Scandinavia's dominant financial group – barred its fund managers from investing in JBS, citing concerns about its environmental record. The key problem is one that Márcio identified early in his career at the company. While JBS can now make sure none of its direct suppliers are committing illegal deforestation, tracking *their* suppliers is a far more complex task.

During its three-year lifespan, an Amazonian bull will typically pass through at least three specialised ranches: born in one, reared to maturity in another and fattened up for slaughter in the last. Comprehensive records on all cattle movement are kept by the agriculture ministry, to guard against outbreaks of disease. The ministry keeps these records strictly secret – and suppliers are unwilling to let JBS have such detailed information on their dealings, for fear of weakening their bargaining position in the ruthlessly competitive cattle trading market.

It was an agonising so-near-yet-so-far situation. All the information required to clean up JBS's supply chain was there, on a government database to which it was forbidden access. Then Márcio and his colleagues hit upon a solution.

In the run-up to Christmas 2017, media all over the world launched into breathless coverage of a geeky innovation that had suddenly become fabulously lucrative. Bitcoin had been developed nearly a decade earlier by a mysterious computer nerd, or team of nerds, who went by the pseudonym Satoshi Nakamoto. Billed as a new form of currency outside the control of governments and central banks, it ran on a technology called blockchain, which promised to store records by means of encryption that was virtually impossible to hack.

The boom in the bitcoin price looked to many like a classic bubble. But the media extravaganza put the underlying technology firmly on the map, and heightened the interest of executives around the world in its potential business uses. For JBS, Márcio realised, blockchain offered a potential answer to its most stubborn problem.

A few days before my visit, Márcio joined the company's top executives to unveil the JBS Green Platform, a blockchain-based system that, it promises, will finally eliminate illegal deforestation from every part of its supply chain. It was billed as a transformational moment for the entire business; even the lifts inside the São Paulo headquarters have their doors emblazoned with the slogan *Juntos pela Amazônia* – Together for the Amazon.

JBS suppliers will be required to register on the new system, which will check their details against the agriculture ministry database to identify their own suppliers, and the suppliers to those suppliers.

Each of those farmers is then checked against the database of violations maintained by Ibama, the environmental agency. If any rancher in his supply chain is on the Ibama blacklist, the supplier will be blocked from selling to JBS. Crucially, the blockchain encryption means the system will only tell JBS whether farmers' supply chains are clean, without letting the company spy on any sensitive commercial information.

It might still take time, Márcio concedes, to get some sceptical ranchers on board. But from 2025, any farm that hasn't signed up to the platform will be barred from doing business with JBS. And while the suppliers it's blocked to date have until now had little trouble finding other buyers for their animals, Márcio is hoping that Brazil's other meat companies will also adopt the new platform, which has been made open for anyone to use.

Hasn't JBS missed an opportunity to gain a competitive edge? By keeping this platform for its own use, I point out, it could claim to have by far the cleanest beef in Brazil. The situation is too serious for that, Márcio says. As long as cattle farming continues to drive deforestation in the Amazon, JBS will be damaged by association. 'This is a problem for our whole industry – for the perception of Brazil itself,' Márcio says.

But even as JBS rushes to strengthen its environmental record and secure the loyalty of customers and investors, yet another threat is coming into view, one that was the stuff of science fiction not long ago. In tech hubs around the world, a new generation of companies are turning out meat that is guaranteed not to be carved from illegally raised animals. It isn't carved from animals at all.

CHAPTER 16

It would be impossible for a global population of 10 billion people to eat the amount of meat typical of diets in North America and Europe and keep within the agreed sustainable development goals for the environment and climate . . . The food system needs to change radically to address these challenges, and a very important part of this will be the adoption of new technologies.

– White Paper on Alternative Proteins, prepared by the Oxford Martin School for the World Economic Forum (January 2019)

HAIFA, ISRAEL

The Sunday early morning train rumbles north between the Carmel Mountains and the Mediterranean Sea, crammed with olive-clad teenage conscripts of both sexes juggling rifles and smartphones, dozing and gossiping through their journey back to the northern frontier, at the end of the Israeli weekend. I get off at Haifa, the city that grew around Israel's finest natural harbour, where countless thousands of Jews disembarked from Europe in the years following the Holocaust. The melting pot of immigrants produced a thriving intellectual environment, driving the growth of world-class research hubs including Haifa's own Technion, a technology-focused institution on a steep hill above the port.

A constant threat of conflict in a hostile neighbourhood pushed Israel to invest heavily in its armed forces, including some of the world's most advanced military research and development, helping to develop the skills and infrastructure for a thriving commercial tech scene. Across this tiny country, young Israelis set to work on technological breakthroughs that could underpin high-growth startups or be sold for handsome sums to Silicon Valley.

Shulamit Levenberg was not among them. Raised in Zichron Yaakov, a small town 15 miles south of Haifa, she dreamed not of lucrative entrepreneurial success, but of throwing light on the workings of life. In the 1990s she won a place to work alongside Robert Langer, a star professor at the Massachusetts Institute of Technology, who was becoming known as the 'Edison of medicine', building animal and human tissue in his lab using stem cells and polymer frameworks. Shulamit proved one of his lab's most brilliant members and went on to become a world-leading scientist in her own right.

We meet in her office at the Technion, where she radiates the intellectual confidence of an academic now established as one of the top minds in her field. Since returning to Israel more than a decade ago, she has achieved a string of medical breakthroughs offering new hope to cardiac patients, including engineered tissue that can start building new blood vessels after being inserted into a damaged heart. Now, she finds herself involved in the country's startup scene, applying her expertise to an altogether different goal: producing tissue not for medical treatment, but for eating.

If it can reach commercial scale, the production of meat through cell engineering looks like a powerful tool in the climate struggle. It could enable the world to meet massive growth in meat demand from countries like China, while dramatically minimising the impact on the climate from methane-belching livestock, and the tracts of rainforest cleared to feed them. And a new source of safe, low-cost protein will help to address rising concerns about how we will feed a soaring global population as traditional agriculture worldwide is hit by myriad adverse impacts of a warming planet.

Shulamit long resisted the temptation to turn her work in this direction. As early as 2005, she received an email from a businessman asking whether she had considered applying her work to the meat market. She demurred. 'We had to focus on the medical applications,' she says now. 'It seemed too early to go in other directions.' She held off even as other researchers dived into the area – notably the Dutch professor Mark Post, who attracted huge attention in 2013 by unveiling the world's first cell-based beefburger.

But, finally, she took the plunge and started applying her technology to meat production. Where other companies were producing food that resembled minced meat – in burger or nugget form – Shulamit's technology, which can produce whole pieces of muscles when applied to medical challenges, could yield entire steaks or chicken breasts. This concept gave birth to Aleph Farms, the most high-profile of a plethora of 'food tech' companies springing up in

Israel, in a new incarnation of the country's long-running startup boom. Its name and logo derived from the first letter of the ancient Hebrew alphabet – a bullock's head – Aleph wants to be the first company in the world to produce whole-muscle meat in the lab and, eventually, in factories.

As Shulamit and I speak on the Technion's modernist campus, one of Aleph's meaty creations is in orbit, nearly 300 miles above the earth, inside the International Space Station. It's just a few weeks since Aleph announced it had become the first company to grow meat in space – small chunks of beef created with Russian 3D bioprinting technology. The breakthrough attracted a burst of excited global media coverage, and will surely be of interest to researchers seeking ways to keep crews nourished on extended space voyages. But the astronaut market will remain limited for the foreseeable future, acknowledges Didier Toubia, Aleph's CEO, as we tuck into a vegetarian lunch of quiche and salad at the company headquarters in Rehovot, a tech hub near Tel Aviv, 60 miles south of Haifa. 'We're focusing,' he says, 'on Earth.'

Born in Paris, Didier got to know Israel through family visits as a child, and through exchange studies at the Technion as a student. When he decided to pursue a career as an entrepreneur, Israel's buzzing startup environment made it the logical place to go, and he made a name for himself in the country's world-leading medical technology sector.

In 2016 he received a call from Jonathan Berger, a tall, square-jawed business school classmate who was now running a food-tech incubator for Strauss, Israel's biggest food company. Intrigued by Mark Post's burger breakthrough, Berger had flown to the Netherlands and spent several hours with the researcher in his Maastricht laboratory. Returning to Israel, Berger had scoured the country's universities for academics whose work was ripe for commercial exploitation, and then its startup scene for a CEO who could bring it to the market. Shulamit and Didier were his dream team.

In a cramped row of labs at Aleph's headquarters, I find the bioreactors where its researchers are growing its meat – flasks about the size of a beer mug, with pipes flowing in and out to regulate the pale reddish liquid inside them. The key objective now is to perfect that fluid, which provides nutrition to the animal cells – finding a way to supercharge the growth process while keeping cost down. It's a matter, says Didier, of tricking the cells into thinking they're inside a cow. The recipe for Aleph's growth medium is a closely guarded secret. So, too, are the design details of the tissue bioreactor created by Aleph's scientists, which they are using, in a separate room closed to outsiders, to hone the structure of their fake steak, a carefully designed blend of muscle and fat arranged on a scaffold of plant tissue.

A year before our meeting, Aleph went public with its first proof of concept. It was the culmination of months spent producing dozens of early versions, thin strips of meat grown in the Rehovot lab. For taste tests, Aleph had brought in Amir Ilan, a well-known Tel Aviv chef, who was unsatisfied by the initial specimens. But towards the end of 2018, as Didier looked on, Amir cooked up one of the lab's creations in his kitchen in northern Tel Aviv and declared himself impressed. Triumphant, Aleph released videos of people at a dinner party tucking into its meat, served by a beaming Amir.

While they were a powerful proof of the Aleph concept, there was still a gulf between those early specimens and the commercial market. By Didier's own admission, the taste still fell far short of quality meat produced the traditional way. Each strip of Aleph's meat was just a couple of inches long – enough to look elegant when presented by Amir, nouvelle cuisine-style, alongside a light garnish, but far too puny to satisfy any serious steak lover.

That will change, Didier says, with the commercial prototype that his team is now creating. This time, the steaks will be far larger, using Shulamit's technology to build a network of blood vessels inside each one, enabling nutrients to permeate into thick muscle tissue. Just as with a normal steak, the new one will be marbled with streaks of fat, giving the juicy texture required by dedicated carnivores.

But making a tasty steak is just the first of many hurdles that Didier faces. He'll need to get clearance in all his target markets – including the US, Europe and major Asian economies – from regulators for whom this is strange and novel territory. Especially in Europe, they have shown a cautious approach towards other innovations such as genetically modified foods.

Then there is the question of cost. Each of those little strips of meat cooked up a year earlier by Chef Amir had a marginal production cost of $50 – for just a few mouthfuls, prohibitive even to well-heeled gourmets. The bulk of that cost comes from the growth medium – the reason why so much effort, at the time of my visit, is focused on that pink liquid. In any case, when Aleph gets its product on the market it will almost certainly still be much more expensive, pound for pound, than traditional beef. Didier is planning to pursue a Tesla-esque strategy: starting by targeting wealthier consumers, attracting them through his product's green credentials and novelty value while costs remain high; then building up economies of scale that will enable him to tackle the mass market with large factories filled with huge steel bioreactors that will look, he says, much like breweries.

Eventually, he insists, Aleph's product should be significantly cheaper than conventional meat. It will take just three weeks to grow each of its steaks from a single cell, compared with the three years taken to grow a cow to slaughter age. And during that time, the amount of nutrition required, per gram of meat produced, will be far lower than for animals reared the traditional way. None of the nutritional energy pumped into Aleph's steaks, after all, needs to go towards growing bones, skin, ears, hooves and internal organs. None is used to power the cow's contemplative strolls around its pasture. None of it gets expelled as methane from the four-stomached bovine digestive system.

But it could be a struggle, I suggest, to attract a mass market that is often nervous about the application of technology to food. Look at all the 'Frankenfood' labels hurled at GM crops – plants that grow in fields just like normal ones, but with a single gene altered by scientists. Entire pieces of flesh taking form in bioreactors, surely, are much closer to Mary Shelley's monstrous vision. It's an objection that Didier is used to fielding, and one that he himself had to grapple with before he took the helm of Aleph. He has overseen a defiant approach to the question of 'naturalness' in the company's branding. '*Meat growers . . . Emulating nature's genius*' read the slogans emblazoned on the wall that greets visitors to Aleph's office. Didier likes to stress the fact that there's no genetic modification involved in the Aleph process, and the ways in which his product is closer to 'natural' meat than the output of modern beef farming. None of the antibiotics, for example, that are pumped into cows to stop them falling sick in their miserable industrial farm stalls. 'It's real meat,' he says. 'The same meat we know. But produced in an improved manner.'

In a few weeks, Didier says, he'll fly to France to address another key audience: the country's politically powerful farmers. In the US, animal farmers have banded together to resist the threat of lab-grown meat, successfully lobbying for legislation in some states stipulating that only flesh taken from a dead animal can be labelled as meat. In his speech to farmers in Rennes, Didier is anticipating a tough reception. 'They want to be reassured that we're not going to kill them,' he smiles. Instead of replacing farmers, Didier says, Aleph wants to incorporate them into its business plan. He outlines a futuristic vision of a European countryside where farms play host to gleaming steel bioreactors. 'Feeding the world is the role of the farmer, not the role of the startups,' he says. This modular, distributed approach would enable food to be produced wherever it's needed – helping to reduce the huge environmental footprint that comes from the transportation of meat. China imports three million tonnes of beef a year, with the lion's share of it sailing more than 10,000 nautical miles from South America – about the least sustainable sourcing imaginable, Didier notes.

For now, the French countryside feels a long way off as Didier's small team of scientists, supported by Shulamit's researchers in Haifa, work flat out towards getting a first product on the market. Part of the urgency is driven by the chance to make an impact on one of the twenty-first century's most serious challenges. But there is another reason to move fast. Aleph Farms faces serious competition.

* * *

Seven thousand miles away in California, Chris Davis and his research team have been confronted by a question that may never have occurred to anyone else in the history of human food production: should they add a shit smell to their meat?

Inside the Redwood City labs of his company Impossible Foods, Chris and I are standing by an apparatus built to separate a substance into its thousands of constituent compounds, then pump these through a tube into a nosepiece a bit like those found on diving masks. The researchers take it in turns to spend half an hour with their noses jammed into the device, cataloguing the aromas coming from the latest specimen of Impossible's plant-based meat products.

By the machine lies a notebook used by the latest sniff tester, who has left a time-coded register of smells that veers from 'sweet floral candy' to 'cheesy butyric acid' to 'burnt apricots' to 'cheesy feet' to 'animals' and back to feet, this time 'musty'. Each of these will be logged and matched with a specific compound. And while most might sound pretty unappetising, together they might add up to what Impossible is searching for: a thick, tangy, undeniably *meaty* aroma that will be enough to satisfy the most hardened carnivore, enough to make them forget that what they're eating comes entirely from plants.

When the researchers put real meat through the sniff test, they often find themselves confronted by the unmistakable smell of animal dung. Even modern abattoirs, fully compliant with food safety regulations, find it impossible to stop microscopic waste particles escaping into the air and circulating through the plant. In 2015, *Consumer Reports* magazine sampled 300 packages of ground beef from 103 food shops in 26 cities across the US. Every single one contained faecal bacteria.

All this presents the Impossible researchers with a dilemma. If their mission is to recreate the aroma of meat in all its mysterious complexity, and if faecal aerosols are part of that inexplicably appetising blend, then should they include in their burgers a replica of the chemical footprint from those gastrointestinal microbes?

The question has been debated internally – but, so far, the answer has been no. Impossible doesn't want to limit itself to making imitations of conventional meat, says Chris, an Englishman with an Oxford doctorate in organic chemistry. It wants to reinvent the whole concept of meat – to persuade consumers to stop defining meat as dead animal flesh and accept the idea that it can be a high-tech product made from sustainable, plant-based components.

Meat, for Chris, is simply 'a consumer good with specific properties that people like'. And when you start seeing meat as a technology product, it makes sense to apply the Silicon Valley framework of relentless improvement. 'You should start thinking of it like a phone, with new features being delivered on an upgrade cycle,' Chris says. 'People should start assuming that every year or two their meat will be cheaper, tastier, more nutritious than it used to be. You should *demand* that from your meat provider.'

It's a vision of an industry so transformed that it's hard to see how traditional meat producers could compete. And for Impossible founder Pat Brown, that's the whole point.

'The first order of business is to crash the US beef industry,' Pat tells me later, wearing a mauve hoodie left unzipped to display his favourite T-shirt: a cow's head in a red circle sliced by a red diagonal line. Outlandish as that target might seem, it's just a stage in Pat's grand plan: the elimination of the whole global animal meat industry, by 2035.

As a breed, startup founders are prone to setting brash, grandiose goals – it's an obvious way to energise a newly assembled team, and to generate buzz among media and investors. But Pat is the first I've encountered whose fundamental drive seems to come, not so much from the thrill of creation, as from a hunger for destruction: that of the industry that he considers the single biggest hazard to our planet.

The son of a CIA agent, Pat had an itinerant childhood before settling in Chicago, where he earned a biochemistry doctorate and a medical degree. After a few years as a paediatrician, he moved to California to start an academic career that would establish him as a leading authority on gene analysis, helping to build smarter ways to diagnose and treat cancer. A tall, bespectacled marathon runner, he was well into his fifties when, during a sabbatical from Stanford University, he started looking for a new problem to attack. He became fixated on the global meat industry, outraged by its continued existence. For one thing, it seemed a stunningly inefficient use of planetary resources. Meat companies present their products as an important source of protein for humanity – but as far as Pat is concerned the sector is a protein destroyer, consuming far more nutrition in animal feed than it supplies in the form of meat. 'If we could just flip that industry off, the global protein supply would actually be higher than it is today,' Pat says. That's before you get to the vast use of increasingly precious fresh water (about 2,400 litres per beefburger), and of land (worldwide, 15 million square miles is used to support animals: quadruple the land area of the United States).

And then there's the methane, belched and farted in huge quantities by a global cow population that has risen, thanks to modern industrial farming, to more than a billion. As I learned from Sergey Zimov in Siberia, methane is a fearsomely potent greenhouse gas, with a warming effect dozens of times stronger than carbon dioxide. But while CO_2 can linger in the atmosphere for centuries, methane breaks down in about a decade. So if we're looking for a fast, effective way to pull the world back from a disastrous climate tipping point, Pat concluded, the obvious place to start is by annihilating the biggest industrial source of methane: meat farming.

Pat was wearily familiar with the insipid meat substitutes on the market, a parade of mushy grey blobs made for committed vegans like him. It hadn't occurred to their designers to capture the interest of carnivores – but that, he realised, would be the key if there were to be any chance of supplanting the meat industry. What if he could decode the molecular chemistry that makes meat so delicious, and then build a nearly identical substance from plants?

Armed with little more than a novel concept and a stellar academic record, Pat secured $9 million in startup funding from an investment firm run by tech billionaire Vinod Khosla. He used the money to assemble a crack team of biochemists, and kicked off one of the strangest reverse engineering projects in the history of California's startup sector.

The secret to making plants taste like meat, Pat suspected from the start, was haem – an iron-rich substance that is found in nearly all living things, but is especially abundant in animals, where it forms the haemoglobin that makes blood red. 'You can take a bowl of vegetable broth, throw haem into it, and now it tastes like beef broth,' Pat says. 'Unmistakably meat.'

In the Impossible labs, I walk through the room where researchers are perfecting the haem production process, in rows of flasks filled with what looks like diluted blood. With their expertise in genetics, it took Pat's team just two years to devise a system to churn out haem in industrial quantities, by inserting a soy gene into the DNA of a yeast strain. Then they set about making their first burger, spending years analysing the chemistry of ground beef and trying to recreate it with a long succession of experimental plant-based recipes.

Toconao, an oasis village in the middle of Chile's Atacama Desert. Local farmers are turning from thirstier crops to wine production, to make better use of dwindling water supplies.

A herder in the Kori district of Ethiopia's Afar region. Herds have been devastated by protracted drought followed by a severe locust infestation.

Said Ugas with his bodyguards in the abandoned town of Mustahil in Ethiopia's Somali region. He has led the relocation of the town's whole population following increasingly severe flooding.

Vandana Shiva, one of the most influential – and controversial – figures in the global debate over genetically modified crops.

Tuvtsengel (right) with his uncle Tseveen in Uvs, Mongolia. He had lost nearly half his 500 animals in the previous winter.

Gombo, a herder in Mongolia's Zavkhan province. He'd been hit by exceptionally deep snowfall followed by temperature swings that caused the top layer of snow to melt then freeze. That formed an ice layer that his animals struggled to penetrate to reach the grass beneath.

I met Ezio when he was burning a large patch of rainforest outside the town of Boca do Acre in Brazil's Amazonas state. He plans to turn the area into cattle pasture.

Awapy Uru-Eu-Wau-Wau (left, with his friend Tebu) leads a group of young indigenous people resisting invasions of their territory by illegal loggers. His brother-in-law was murdered a few months before my visit.

Workers at the Mainetec factory in Mackay, Queensland. Mainetec founder Brett Hampson told me that opponents of Australia's coal industry overlook its economic contribution.

AGL manager Paul Barrand at the Loy Yang mine in Latrobe. AGL – Australia's largest electricity provider – and its partners want to use coal from the mine to make hydrogen for Japan's new energy market.

Iceland's Hellisheiði geothermal plant, where Swiss company Climeworks and local partner Carbfix are pioneering a system to suck carbon dioxide from the air and turn it into underground limestone.

Yisha He at her office in Shanghai – one of an increasingly powerful set of entrepreneurs in China's solar energy industry.

He Xiaopeng with the Xpeng P7 in Guangzhou. Having made a fortune in China's mobile internet sector, the billionaire now leads one of the country's biggest electric car makers.

Howard Schmidt at the Quidnet test site in Texas. The company wants to deploy shale drilling technology to store renewable energy, using water and underground rock formations.

Mangovo holds cobalt ore in a mine chamber under Kasulo. Over 100,000 Congolese people work at these 'informal' mines, digging the cobalt that is a vital ingredient in most electric car batteries.

Monique Swila, one of many small entrepreneurs in the Congolese mining sector, at her cobalt mine in Kasulo.

The company's first product, a burger made from wheat, potato, coconut oil, various natural flavourings and gums and lashings of haem, hit the market in 2016 – limited, due to its still exorbitant cost and small production volumes, to a handful of upmarket restaurants on both coasts of the US, where it got strong reviews for its convincing flavour and haem-fuelled ability to 'bleed' on the plate. But it was three years later with the launch of the Impossible Burger 2.0 – at Las Vegas's Consumer Electronics Show, an annual gala for the global tech industry – that momentum began to build. Now built on soy instead of wheat, the new model tasted much meatier than its predecessor as well as being lower in fat, less crumbly in a frying pan and – crucially – much cheaper to produce.

'That is insane! That is amazing!' the fire-breathing right-wing provocateur Glenn Beck marvelled on his radio show, after incorrectly identifying the Impossible Burger as real meat in a blind taste test. 'I could go vegan! And I'm a rancher!' A few months after the launch, Burger King rolled out the Impossible Whopper – priced at just a dollar more than the cow-based standard – and other fast-food chains lined up to join it. As Impossible struggled to ramp up supply, employees faced ballooning workloads and morale came under heavy strain. 'The organization is eating itself alive,' one employee wrote on the job review site Glassdoor. 'The CEO has good intentions (and is a true scientific genius) but is a terrible business leader,' wrote another. But by the time of my visit to Impossible, its products are on sale at more than 30,000 restaurants and nearly 20,000 food shops across the US; among the latest recruits is Starbucks, which has just started selling the Impossible Breakfast Sandwich, made with plant-based sausage.

With startling speed, Pat's vision of an explosion in demand for plant-based meat has gone from eccentric prophecy to Wall Street consensus. Even his nemeses in the animal meat industry are now chasing a piece of the action. Tyson Foods, the biggest US meat producer, is ploughing millions into plant-based products while rebranding itself not as a meat company but as an all-purpose 'protein leader'. In Brazil, JBS has set up a plant-based team under food scientist Renata Nascimento, with whom I shared a meal of soy-based nuggets and kibbeh in São Paulo. I found that lunch surprisingly tasty, but Pat dismisses the plant-based efforts of the incumbent meat groups as a publicity stunt. 'They're not serious about making products that will compete against meat from animals. It doesn't make any sense. They have so much sunk cost in there. It's their whole industry!'

But there's a growing crowd of rival startups set up specifically to make plant-based meat – notably Los Angeles-based Beyond Meat, which makes much of the fact that, unlike Impossible, its products are free of genetically modified ingredients. (Pat comes down firmly on Usha Zehr's side of this debate, saying it is 'absolutely not an issue' for 'anyone who is scientifically savvy' – a category that may not cover his entire target market of several billion meat eaters.)

And then there are the cell-based companies like Aleph Farms, whose work Pat seems to consider borderline absurd. 'If they succeed, I'll be their biggest fan, but they won't,' he says. They'll never manage to reduce their costs enough to make economic sense, he insists. And even if they do, he gives them no chance of competing on flavour and nutrition with Impossible's team of scientists, who are pursuing a perpetual upgrade cycle with the entire plant kingdom at their disposal. 'It's as if two hundred years ago someone said: "You know, this whole horse thing isn't working out too well, the streets of New York are filling up with horse shit. So here's my genius idea. We're going to grow horse muscles from stem cells, and hook them up to gears and pulleys." You've just missed the whole opportunity!'

Still, Pat insists, the goal is not to wipe out competitors – except the ones selling dead animal bits. It might turn out, he muses, that Impossible ends up with only a tenth of the global protein market. 'That market is projected to be worth $3 trillion in 2030. So if we wind up with a tenth of that, our investors have nothing to complain about.'

Pat's downbeat scenario, I realise, appears to be a business with annual turnover of $300 billion – more than the latest figure for Apple, Amazon or any US company bar Walmart. He's taking for granted an industrial disruption at a speed and scale rarely seen in economic history. But as we speak, a still more powerful transformation is already well under way: that of the energy systems that power modern civilisation.

PART SIX

Fossils

CHAPTER 17

I think by 2020, if oil stops we can survive . . .
The vision is not a dream, it's a reality that will come true.

– Mohammad bin Salman Al-Saud, then deputy crown prince of Saudi Arabia, 26 April 2016

RIYADH, SAUDI ARABIA

At his cavernous office within a gleaming government complex in central Riyadh, Prince Abdulaziz bin Salman Al-Saud is reflecting on the sad fate of the video rental chain Blockbuster.

'What happened to it? Gone!' says the prince, puffing on an e-cigarette and clad in a flowing white thobe. 'Why? Because technology took over.'

The energy sector needs to brace itself for similar upheaval, says the prince, who as Saudi Arabia's energy minister is the most powerful man in the global oil business. Over nearly four decades in the industry, he's watched concerns about carbon emissions go from a sideshow to an existential threat. Now the world's hydrocarbon exporters need to find ways to tackle the environmental impact of their oil and gas and find new low-emission revenue streams. Those who fail to do so, he warns, 'will render themselves obsolete', no less than a purveyor of Hollywood films on tape.

Perhaps no country in history has ever owed so much to a single commodity as Saudi Arabia owes to oil. When the prince's grandfather, also named Abdulaziz, unified this territory after decades of tribal warfare in 1932, nearly all his people were illiterate, many of them roaming the desert with camels as they had for centuries. Then the country's enormous oil reserves were discovered, and Saudi Arabia rose over the following decades to become the world's biggest exporter. At meetings of OPEC, the club of major crude producers, Prince Abdulaziz is by far the biggest fish in the room, able to hike or crash global petroleum prices with a mere tweak of the kingdom's output. By channelling the prodigious oil revenues into perks for the country's population – now mostly living in cities like the sprawling metropolis of modern Riyadh – the Al-Saud family has been able to keep unchallenged autocratic power for nearly a century, while much of their region has lapsed into military coups and civil wars.

Now, the formula that underpins their rule has been called into question. Until recently, global oil demand looked set for inexorable growth. But with the dramatic increase in concern about climate change, and the accelerating rise of electric cars, many analysts say it will soon enter a steady decline, with prices set to be squeezed as producers try to make the most of a dwindling market.

The kingdom's heavy reliance on oil revenue has given it an obvious incentive to resist this trend. It's notorious in the environmental movement for having worked to block ambitious targets at global climate talks – an unfair characterisation, the prince claims. 'Actually, you are bullied at these things,' he says, accusing Western leaders of unrealistic promises that threaten to destabilise the global economy. 'A politician is a politician: he would sell a pipe dream, because he's not there in 2050, so he's fine to throw the problem somewhere down the road.'

Still, he insists, Saudi Arabia is no longer trying to hold back the transition to a low-emissions energy system – but is instead rushing to ready itself for the new era, mindful of the shift in momentum that is driving those bearish forecasts for global oil demand. 'You have to ask yourself how resilient you will be,' Prince Abdulaziz tells me, 'if these trajectories come to fruition.'

Since his father Salman came to the throne in 2015, this country has embarked on a breathless mission to build that resilience – both by equipping the huge energy sector to compete in the low-carbon era, and by developing new industries to offer jobs and prosperity to a youthful population. Already 79 when he took the throne, King Salman has entrusted that drive to his favourite son: Mohammad bin Salman, widely known as MBS, who in his mid-thirties is now the uncontested de facto ruler of Saudi Arabia, and one of the world's most controversial leaders.

Soon after becoming crown prince, MBS stamped his mark on the kingdom by turning Riyadh's Ritz-Carlton hotel into a prison for dozens of Saudi Arabia's richest men, who were forced to hand over tens of billions in allegedly ill-gotten gains. 'It started in a clumsy way, because it was shock and awe,' concedes Prince Abdulaziz, an older half-brother whom MBS has named as a role model during his youth.

But that's the kind of drive and energy that's needed, he tells me during a three-hour discussion over innumerable cups of Arabian coffee, to shake the country out of the inertia that could cripple its long-term future if it fails to move with the times. 'Whatever is achievable in a week, he wants achieved in the next hour. It's harsh and hard. But it's making the machine work harder,' Prince Abdulaziz says. 'Big, huge ambition.'

* * *

Shortly before my visit to the kingdom, MBS put out a video in which he revealed new details of his most extravagant project. 'By 2050, one billion people will be displaced due to rising CO_2 emissions and sea levels,' the stocky, full-bearded young prince predicted, jabbing the air with his right hand. 'Why should we sacrifice nature for the sake of development?'

The answer, he went on, was something called The Line: a city quite unlike any other the world has yet seen, with 'zero cars, zero streets, zero carbon emissions'. Instead, this pencil-shaped settlement of a million people will be built on a 100-mile-long invisible underground infrastructure system, and powered entirely by wind and solar energy. 'We need to transform the concept of a conventional city into a futuristic one,' MBS declared.

The Line is to be the centrepiece of Neom, a 10,000-square-mile development zone in Saudi Arabia's far northwest, intended to transform the country's image both for the wider world and for Saudis themselves. Instead of a sclerotic fossil-fuel economy, with conservative rulers stifling the spark of innovation, MBS wants Neom to be the cradle of a high-tech, zero-carbon model in which Saudi Arabia will blaze a trail for the rest of humanity. On his $500 million yacht, moored in the Red Sea near the Neom site, MBS's imagination was fired by visits from a stream of highly paid Western consultants. This would be by far the world's greenest city, with all its electricity provided by zero-emission sources, and fossil-fuel vehicles banned from entering. That was just the start of the vision, according to planning documents leaked to the *Wall Street Journal*. Neom would have a giant artificial moon, and the sand on its beaches would glow in the dark. It would be full of robots – some cleaning homes, others in the form of dinosaurs roaming a techno-Jurassic Park, still others sparring as gladiators in cage fights.

When these details reached Ali Shihabi, a veteran Saudi businessman, he grew concerned. This was just the latest in a long line of grand projects in the kingdom to build new desert cities or flashy business zones. Most of the others had proved expensive flops. Neom was by far the most ambitious – and, with a planned price tag of $500 billion, dozens of times more costly than any of the previous schemes. The son of an ambassador, Ali had got to know the royal family over the years, and he visited MBS to outline his concerns. 'I said, "I think it sounds like pie in the sky,"' Ali recalls. 'Everybody says he's not open to criticism. But he said, "Well, put your money where your mouth is and join the advisory board." And I said, "Sure."'

Since becoming directly involved in the Neom project, Ali has come to see some substance beneath the overhyped marketing. For one thing, the site is undeniably blessed with exceptional potential for renewable energy generation. It's in the sunniest region of one of the planet's sunniest countries, rivalled only by the world's great deserts. There are strong winds, too, rolling in from the Red Sea. Most intriguing for Ali is the plan to build huge, dome-shaped, next-generation desalination plants to provide the new settlement with fresh water. If it works, the technology could prove valuable for this arid country and the wider Gulf region, as it copes with growing pressure on water supplies amid the shift to a hotter climate. Already, Saudi Arabia is heavily reliant on its coastal desalination plants, which currently consume large amounts of fossil fuel energy and churn out concentrated brine that is toxic to marine life. Instead this plant, the first built on an untried new model developed by a British startup, will use the sun's heat to boil water from the Red Sea, and the salt residue will be safely disposed of or put to industrial use.

'Would I have launched Neom? No,' says 61-year-old Ali. 'But I have the cynicism of an older chap. And that's why I think there is a very refreshing thing to young leadership. Yes, they may be a bit naive. But they have such enthusiasm, such drive. Who am I to say it won't work?'

The world's most powerful millennial, set to rule the Middle East's leading power for the next half-century, MBS is part of a swollen youth demographic in a nation where three-quarters of people are under forty. He's boosted his popularity among his generation by sweeping away many of the country's most infamous strictures. Women can now drive, once-illegal cinemas are proliferating across the kingdom, and the feared religious police have vanished from the streets.

To many outsiders, however, MBS is known not as a green-minded reformer but as a suspected killer. In October 2018, just as MBS was revving up a global publicity drive around Neom, news broke of the murder and dismemberment in Istanbul's Saudi consulate of Jamal Khashoggi, a *Washington Post* columnist who had been the prince's most prominent Saudi critic. MBS accepted ultimate responsibility for the killing while denying any personal involvement. The stellar list of advisers he had assembled for Neom, from ex-US energy secretary Ernest Moniz to former European Union commissioner Neelie Kroes, jumped ship en masse.

MBS's critics allege that the murder reflected a broader malaise, that the talk of zero-carbon innovation is a smokescreen for a ruthless approach to dissent and human rights. Even as the ban on female drivers was dropped, women who had campaigned for that move were being thrown in prison. The rate of executions has surged under the new leadership, which is using capital punishment as a 'political weapon against dissidents' according to Amnesty International. At Neom, amid the talk of living in harmony with nature, 20,000 members of the Huwaitat tribe will be forced from their homeland to make way for the development. The government has promised that they will be well compensated and given new homes better than their old ones. But some in the tribe have publicly resisted the move, including Abdul Rahim al-Hwaiti, who condemned the Neom scheme in videos he posted online. He was shot dead by Saudi security forces in 2020. Officials said he fired first.

Ali says such incidents need to be seen in the context of a direly needed overhaul of an economy and society that had been slowly drifting toward crisis. 'It's a car driving down a highway, that used to be going at thirty miles an hour and now it's going at a hundred and twenty. It's going to hit a few holes, maybe it runs over a few people, but it's going in the right direction.' If MBS can confound his critics and build a new, post-oil economy, investing the petroleum wealth wisely to seed new industries, exploiting what could be the lowest-cost solar energy in the world – then, Ali says, there will be no shortage of people willing to do business with and invest in the kingdom. 'Capital goes where it makes money,' he says. 'If the pools of capital see an opportunity to make money in Saudi Arabia, they're not going to give a damn about Jamal Khashoggi.'

* * *

The American prospectors had spent five years fruitlessly wandering the sands of Saudi Arabia's eastern desert when, in the spring of

1938, they hit the jackpot. At a site called Dhahran they discovered what would prove the most lucrative field on earth, an abundant expanse of oil whose shallow depth made it superbly cheap to extract.

Today, Dhahran is the most important place in the global oil business. It's the headquarters of Aramco, the company that grew out of the old American operation and is now owned by the Saudi state. Aramco pumps over twelve million barrels of oil and gas per day: more than ExxonMobil, Chevron, BP and Shell combined. When it floated a small fraction of its stock in 2019 – part of an MBS push to raise funds for his new investments – the share price made Aramco the world's most valuable company, with a market capitalisation of two trillion dollars. While MBS ramps up efforts to diversify the economy away from energy in the coming decades, he'll be relying on Aramco to keep the revenue flowing in to support that drive, in an energy sector that's set for an era of unprecedented disruption.

'Anybody who says that oil doesn't have a future doesn't understand where all the products we have come from,' Ahmad Al-Khowaiter, Aramco's chief technology officer, tells me in a tower at the company's sprawling Dhahran base, which is still laid out like a small US town, decades after the Americans handed over control. For one thing, Ahmad points out, there's the huge range of plastics and petrochemicals that account for more than a tenth of world oil demand. And when it comes to transport, petroleum's demise is still at least several decades away, even under the most aggressive forecasts. But as oil demand begins to decline – a long-term trend that, according to some analysts, may already have begun amid the disruption of the coronavirus pandemic – the competition among producers looks set to become more frenzied. And if, as is widely expected, importers like the European Union start putting a carbon tax on shipments, then oil producers with the best – or, rather, the least bad – environmental performance will gain a commercial edge.

Ahmad's job is to ensure that Aramco stays ahead of the pack.

It's a sign of how far the company's priorities have changed that his CTO position didn't even exist until 2013. 'Technology was seen in the company as more of a luxury than a necessity,' says Ahmad, who joined Aramco as a young engineer in the early 1980s. 'Today it's the opposite. Without technology, we won't succeed.'

When you consider the total carbon footprint of the petrol you burn in your car, up to 40 per cent of it is incurred before the stuff even reaches the pump – from the energy used and gas given off when the oil is pumped from the ground, and then when it's refined and transported. Thanks to the natural shallowness and high purity of its oil deposits, Saudi Arabia already has the lowest emissions from oil extraction of any major producer, according to a 2018 study in *Science*. Now Ahmad's team is chasing innovations to make it even lower. They're pursuing cheaper and more effective ways to capture and store the gas released during oil extraction. They've deployed microscopic tracking particles to monitor oil flows and are deploying artificial intelligence and sensors in a push towards über-efficient autonomous oil fields. In a low-carbon economy, Ahmad says, 'the oil most likely to be used will be the one that is the most economic in those conditions – and that's the one with the lowest emissions'. That's bad news for the sector's most heavily polluting producers, in places like Canada and Venezuela, who release enormous quantities of carbon as they retrieve oil from oozing tar sands. But Aramco, Ahmad promises, 'will be the last man standing'.

Cleaning up its core business is just one part of Ahmad's mandate at Aramco. He's also exploring ways that it can use technology to carve out a place in the new energy sector – notably in hydrogen, which has been generating a surge of international enthusiasm in recent years. At the nearby port of Jubail a few months before my visit, a tanker set sail for Japan loaded with 40 tonnes of hydrogen-rich, carbon-free ammonia, produced by an Aramco subsidiary using natural gas from the eastern oil fields. Japan has put hydrogen, which releases only water vapour when burned, at the heart of its future energy strategy. JERA, the country's biggest electricity company, says it will convert all its coal-burning power plants to ammonia by 2050. Japan has been the loudest cheerleader for hydrogen fuel-cell vehicles which, with their light weight, rapid charging and long range, could be a better option than batteries for trucks and other commercial vehicles.

The EU, too, is targeting a massive expansion of hydrogen infrastructure as part of a new green energy push. It wants to convert heating systems across the continent to hydrogen, and use the gas for energy storage to reinforce an electric grid based on fluctuating flows of solar and wind power. Across a wide range of industries, policymakers are hoping hydrogen will be a means to tackle some of the thorniest remaining problems in the push for a low-carbon economy – for example, replacing coal in the process of making steel from iron.

Hydrogen could be a major new avenue as Aramco fights to keep its leading role in the global energy market. Inevitably, when hydrogen is produced from Aramco's bountiful natural gas reserves, carbon dioxide is given off. But if those waste emissions are captured and pumped into underground reservoirs, then the so-called 'blue hydrogen' should still be seen as a climate-friendly fuel, Ahmad argues. That suggestion is fiercely resisted by many in the environmental movement, who see it as a cynical attempt to carve out a place for fossil fuels in the new energy economy. Their concerns are buttressed by recent scientific research suggesting that blue hydrogen could in fact have a very substantial carbon footprint, thanks to leaks that occur when the natural gas is extracted. It's a needless distraction, opponents say, from what should be the real focus: 'green hydrogen', produced from water without any carbon emissions at all, using renewable electricity.

But on that front too, Saudi Arabia is on the move. On the other side of the country at Neom, MBS has approved one of the world's biggest green hydrogen plants, a $5 billion project due to start operations in 2025. So far, the electrolysis process for green hydrogen has been much more expensive than the natural gas method that Aramco is pursuing. But analysts expect technology advances to bring the cost down fast. And with its abundant solar power resources, the Neom project's backers hope it could produce the cheapest green hydrogen on earth. Whichever colour of hydrogen emerges triumphant, Ahmad says, 'you couldn't have a better situation than Saudi Arabia does when it comes to a hydrogen economy'.

For Prince Abdulaziz in Riyadh, it all reminds him of the greyhound racing tracks he visited on boyhood trips to London, where he learned about 'each-way' wagers, hedging bets between two competing dogs. Even if oil demand is set to slide, sending prices into a slump, Saudi Arabia – with far lower production costs and emissions than most rivals – will stay in the market as other producers drop out one by one. And in the shift to new forms of energy, the prince insists, the country is no less well placed. 'You want renewables? We'll kill you with renewables! You want hydrogen? We'll kill you with hydrogen!' However things turn out in this sector over the coming decades, and even as MBS races to find new growth drivers, the kingdom has no intention of surrendering its crown as a global energy superpower. But for other players in the troubled fossil fuel economy, the end may come much sooner.

CHAPTER 18

These mines have got, you know, ten, twenty, thirty years to run . . . What's important is that we continue to extract and get the value from the opportunity and wealth that's there, that really benefits the rest of this country.

– Scott Morrison, prime minister of Australia, 20 January 2021

BRISBANE, AUSTRALIA

For a man just declared bankrupt, Adrian Burragubba seems to be giving remarkably little thought to his financial situation. It's less than a month since the Brisbane Federal Court delivered this latest blow in his battle against the plan by Adani, a giant Indian conglomerate, to develop a huge new coal mine on his people's ancestral territory. After Adrian failed in a series of legal challenges, citing the land's spiritual importance to his Aboriginal Wangan and Jagalingou community, Adani successfully pursued him for legal costs of 600,000 Australian dollars. 'Activists who use the courts for lawfare should be held to account,' the company said.

Sitting in the Brisbane office of one of his supporters, Adrian dismisses Adani's action as an exercise in futility – for the simple reason that he has no assets worth mentioning. Powerfully built with a white goatee, he's spent his life playing the didgeridoo, at community events or busking in malls. It has preserved the music of his clan but offers minimal financial reward. 'All I own,' Adrian says, 'is my knowledge, my culture and my heritage.'

When Captain James Cook's account of his arrival in Australia was published, his editor quietly cut a section presumably deemed unsuitable for British readers of the time. The indigenous peoples of Australia 'are far more happier than we Europeans,' Cook had written in 1770. 'The earth and Sea of their own accord furnishes them with all things necessary for Life.' Australian Aboriginal communities had not pursued agriculture – which provided Europeans and Asians with huge quantities of carbohydrates to feed large, settled populations, at the cost of vitamin deficiency and arduous labour. While European peasants were sweating over fields of wheat and corn, Australian hunter-gatherers were developing what has come to be known as the Great Tradition, with legends and spiritual practices woven around features of the rugged landscape. But to Cook and those who followed him, the fact that the indigenous Australians were not 'working' the land meant they had no claim to it.

The late eighteenth century ushered in the rapid destruction of the Aboriginal way of life. As it happened, it was the start of massive changes for those European peasants, too, and indeed for nearly everyone else in the world. While Britain's sailors were exploring the southern seas, its inventors were kicking off the Industrial Revolution by harnessing the power of coal, radically transforming the shape of industry and transport. Until then, the global economic growth rate had been close to zero. A typical European of 1500, transported forward to 1700, would not have noticed a great difference in her village's standard of living. It was the exploitation of coal, and then of other fossil fuels, that underlay the economic explosion of the past two centuries, a boom without any sort of precedent in human history.

Coal may have fallen out of use in trains and ships, but it's still a massive part of the global energy mix, accounting for more than a third of the world's electricity generation. And while Britain's historic coal mines are now nearly all abandoned, Australia's are still going strong. It exports more coal than any other country, nearly 400 million tonnes a year. Altogether, mining accounts for more than a tenth of gross domestic product, the highest share in any developed nation. The exploitation of its mineral wealth has helped make Australia one of the world's richest nations, as it supplied resources for the surging global economy of recent decades.

But the full cost of that economy's hydrocarbon foundations is now becoming clear, with stark implications for an already hot and arid nation. During my stay in Australia, more than a hundred wildfires rage in the eastern states of Queensland and New South

Wales. A Queensland official calls this unprecedented for the early spring period: 'an omen, if you will'. He's right: this will prove the worst fire season in national history, incinerating 40,000 square miles of land and killing 33 people to round off the hottest, driest year in Australian records. And as such disasters drive momentum for climate action around the world, the prime target is the coal burned in thermal power plants, by far the most polluting major fossil fuel.

When he learned of Adani's plan to build Australia's biggest coal mine in Queensland, a state already being ravaged by the effects of climate change, it struck Adrian as a grim absurdity, the extractive mentality that had dispossessed his ancestors taken to a pathological extreme. The project would be one of the world's biggest mines producing thermal coal, and the first mine in the vast Galilee Basin, one of the world's largest untapped coal reserves. 'They think they can make large wads of money and then get out of it before the planet cooks,' Adrian says of his country's leaders, his words coming in emphatic bursts like the riffs of a jazz trumpeter. 'It doesn't make any common sense.'

The Adani mine has become an iconic flashpoint in one of this country's most politically fraught debates. Should Australia start bringing down the curtain on one of its most iconic industries – or rush to extract as much value as possible from the thermal coal industry, while it's still there?

* * *

The sparks rush in a torrent towards the welder's reinforced mask, its screen darkened against the fierce light of melting steel. Behind him, dwarfing the workers moving among them, are two rows of 'buckets' – the massive metal scoops used by mechanical excavators at mine sites to grab tonnes of coal in a single sweep. Inside the workshop, the size of an aircraft hangar, the air howls and rattles as the buckets are worked into shape, their sections sealed together by temperatures well over a thousand degrees.

Square-jawed, his hair greying but styled in a trendy undercut, Brett Hampson leads me through the factory floor towards his lightest and strongest bucket to date. Hulk – the teeth on its rim painted a radioactive lime-green to match its name – is Brett's newest offering to the booming coal mines of Queensland, which have driven the growth of Mainetec, the entrepreneurial venture that has consumed the past nine years of his life.

Brett's factory is in the middle of a sprawling industrial park in Mackay, a port town of some 40,000 people on Australia's north-eastern coast. A short boat ride out from Mackay's harbour, the Great Barrier Reef is being ravaged by the effects of rising sea temperatures, which have killed more than half of its coral since 1995. The only hope for the reef – which has created a significant number of jobs in the local tourist industry – is a dramatic change of course in the global energy system, cutting back radically on the use of fossil fuels, especially coal. But that could have painful implications for the still larger number of townspeople who depend for their livelihoods on Queensland's huge coal mining sector, working either in the mines or in the many businesses, like Brett's, that have grown around them.

So far, coal has trumped coral. In Australia's last general election, the constituency containing Mackay was one of many Queensland coal communities that swung strongly behind the conservative coalition government, which had made full-throated pledges to protect the industry. Locals had been infuriated by the arrival of a caravan of environmentalists protesting against the Adani mine, and were suspicious of the opposition Labor Party, which had promised to pursue green growth. Some liberals in Australia were so outraged by the election results in Queensland that they hit social media to call for Quexit – the expulsion from the Australian commonwealth of its most regressive state. For businessmen in Mackay like Brett, that was typical of what they see as an irrational demonisation of their industry, a longstanding contributor to the national prosperity that liberal Australians enjoy even as they hurl barbs at coal. 'I feel,' he says, 'like the world's gone mad.'

But whatever happens with the debate inside Australia, the long-term global outlook for coal demand looks anaemic. As technological advances drive down the cost of renewable energy, there will be ever less need for this prodigiously polluting fuel. By 2050, according to forecasts from BP, global coal demand will fall by something between 25 and 90 per cent. Most of what's left will be ageing plants in the Asian markets on which Australian coal exporters are pinning their hopes – but as they too shift to renewables, China and India will be able to meet the vast bulk of their lingering coal needs from their own huge domestic reserves.

'There will be a famine mentality creeping into this industry soon,' shadow energy minister Mark Butler tells me when we meet in a glass-walled office tower in downtown Sydney. Had Australia's last general election played out as the pollsters forecast, Butler would now be a central player in a new government, implementing one of the most aggressive climate change responses of any major economy. On the campaign trail, his Labor Party promised to reduce the country's carbon emissions by well over a third in just eleven years. This was to be Australia's climate election, some predicted, sweeping away the recalcitrant conservative government of Scott Morrison, who had once brandished a lump of coal on the Senate floor, crying: 'Don't be afraid!' But the prime minister drew blood with his warnings of the risks that Labor's plans posed to the economy and a reliable electricity supply, and he returned to power with an enlarged majority.

In Australia, perhaps more than in any other country, climate and energy policy has become 'the defining thing between right and left,' Butler says – a culture war between green-minded urban liberals and conservatives standing up for hard-working miners. In recent years, the latter group has clearly had the edge. But as climate change impacts begin to hit home in Australia, from searing droughts and wildfires to the devastation of the Great Barrier Reef, Butler is convinced that public opinion is finally starting to shift. And beyond environmental concerns, he says, a managed decline of coal mining in Australia is simply a matter of economic common sense, in response to obvious international trends. 'There's a whole body of opinion that says, "That's bullshit, India's gonna go gangbusters on buying our coal." But it's not,' Butler says. Instead of trying to prop up the slowly dying industry, he argues, the government needs to start preparing the communities that rely on it for life after coal.

Latrobe, a sleepy community of 70,000 people in the southern state of Victoria, is already wrestling with that challenge. Tens of millions of years ago, this area was filled with a swampy mass of peat, which was compressed over geological ages to form enormous swathes of lignite, or brown coal, starting just below the ground and stretching far into the earth. From the early twentieth century, Latrobe's low-cost coal mines fed local power stations that at their height generated 85 per cent of the electricity used in all Victoria. Thousands of jobs were created and passed down through the generations – and the coal reserves looked sufficient to keep the mines going for centuries to come.

In the crisp chill of a late winter afternoon, I stand on the rim of the Loy Yang mine, a gargantuan gash in the earth whose steep slopes, the height of skyscrapers, are terraced like rice farms on Chinese hillsides, the legacy of successive phases of digging over decades. Far below on the mine floor, water has gathered in oblong pools, near which yellow mechanical excavators are at work, rendered toylike and silent by the distance. On the opposite side of the pit, two miles to the south, soar slender towers exhaling gentle streams of incinerated coal, alongside squat concave sisters emitting steam.

But Latrobe's flagship industry is disintegrating fast. While brown coal's peat bog origins mean it's close to the surface and easy to dig up, the high water content means larger quantities of it have to be burned, with greater volumes of carbon emitted, compared with the coal from places like Queensland. As the coal industry's environmental impact comes under political fire, lignite has been receiving the fiercest volleys of all.

One by one, Latrobe's power stations have started shutting down, with their replacement now out of the question. Two of them closed in the five years before my visit. The next will follow in 2028, with its operator saying it will replace the plant – which employs about five hundred full-time staff and several hundred more contractors – with a giant battery to support the region's shift to intermittent solar and wind plants. That will leave just the two power stations at Loy Yang – and their owners are coming under pressure to close them, too, with profits sliding thanks to the boom in low-cost renewables, and public concern growing about the plants' huge carbon footprint.

'We've got too many eggs in the energy basket,' says Latrobe's mayor Graeme Middlemiss, a gruffly cheery figure in his sixties. His grandfather came in 1921 to work on the mine that would fuel the area's first power station complex, where Middlemiss's father went on to work as a mechanical fitter. Middlemiss himself spent his career as an operator at one of the Loy Yang stations, opened in 1984 with state-of-the-art technology that granted him a working life, he fondly recalls, much like Homer Simpson's.

But with Latrobe now haemorrhaging coal jobs, Middlemiss is struggling to find ways to replace them. The workers employed at the mines and power stations are just the start of the coal sector's significance to the valley, Middlemiss tells me. Thousands more jobs rely on providing goods and services to those operations, or on custom from their well-paid employees. 'Our whole economy is skewed in that direction,' Middlemiss says.

He's hoping Latrobe's historical strength in energy could bring some opportunities in its new era. He's been in talks with a Chinese state-owned company that's interested in setting up a workshop in Latrobe to refurbish wind turbines. But it will take a huge number of such projects to replace the jobs bleeding out of the coal industry. 'That operation might employ twenty-five people,' Middlemiss says. 'You go, "Yes – twenty-five jobs! Now only two thousand three hundred and fifty to go!"'

To one side of the power station at Loy Yang, behind a locked gate and bedecked with orange flags, I glimpse the foundations of an installation that offers a glimmer of hope for Latrobe's moribund coal industry. As Japan tries to ramp up its hydrogen economy, a Japanese–Australian joint venture is seeking a piece of the action, through a $390 million project to produce hydrogen in a chemical process using coal from the Loy Yang mine. Despite its huge cost, the experimental plant will run for only a year and produce just 3 tonnes of hydrogen – an amount that would be exhausted by filling up 600 of Toyota's new Mirai fuel-cell cars. With the state and national governments together contributing a large chunk of the investment cost, critics say it is yet another example of Australian politicians' irrational attachment to coal, pumping taxpayers' money into propping up a dying sector instead of ramping up investment in solar and wind.

Middlemiss recently went to the nearby Port of Hastings for the start of work on the plant that will prepare the hydrogen from Loy Yang for transportation to Japan. He was confronted by placard-wielding opponents of the project, who gave short shrift to his pleas to give the technology a chance, his protestations that coal can still be of use in the new economy. Now in his last few months as mayor, Middlemiss seems tired of battling the criticism of his hometown's cornerstone industry. Unless the hydrogen project can defy the odds and reach commercial scale, Latrobe's vast mineral reserves, once a seeming guarantee of perpetual prosperity, are on course to be rendered economically useless, a mere geological curiosity. All the bullish talk from the prime minister is doing little to slow the disappearance of the coal power industry that gave birth to this town. 'Our community is coming to accept,' Middlemiss says, 'that all that is a thing of the past.'

* * *

When Liverpool schoolgirl Fiona Wild was figuring out what to do with her life in the late 1980s, the world – or, at least, its politicians and media – was just starting to wake up to the scale of the threat posed by climate change. Like many other teenagers, Fiona was shocked and energised by what she learned about the dangers of global warming. But while others resolved to join environmental groups and campaign against the fuel companies they blamed for the crisis, Fiona decided to pursue a different tack. Instead of doing battle with the resources sector, she wanted to change it from within.

We meet at the downtown Melbourne headquarters of BHP, Australia's largest company and the world's biggest mining group – a huge glass structure thrown up behind a century-old brick façade with delicate cream cornices. Three decades on from her wake-up moment, Fiona is precisely where she set out to be: head of climate strategy at a behemoth of the carbon-belching global economy. 'I honestly think we can effect more change here, from a climate perspective, than we could in almost any other organisation,' Fiona says.

BHP extracts and sells more than $40 billion worth of commodities a year, and the carbon emissions from their use are estimated at well over 500 million tonnes: higher than the national figures for all but the biggest countries. And as BHP's leaders come under pressure to do something about this massive carbon footprint, Fiona's team is now at the centre of the company's long-term strategy.

Two months before our meeting, chief executive Andrew Mackenzie announced that BHP would start investing in reducing not only the emissions produced from its own direct operations, but also those released when its coal, iron ore and other minerals were used by its customers: an amount nearly forty times greater. Fiona's eyes squeeze in impish glee as she describes the response of rivals in the sector. 'Some of these other players have been like, "Oh BHP you crazy fools, what are you doing? Because if you go there, then we have to go there too." And, yes, they do.'

The wry smile returns when she recounts the dismayed response of the head of BHP's thermal coal division after her team issued a report warning the unit – which produces coal for power plants – would be a 'loser' under a rapid shift towards cleaner fuels. When we meet, BHP is quietly undertaking an internal review of that operation, which later results in a decision to sell its thermal coal mines and wind down the business. It's part of a broader shift in the company to pivot towards resources, like nickel and copper, that have a bright future in the age of renewable energy and electric cars.

But BHP still has a huge portfolio of metallurgical coal, used in steelmaking, as well as offshore oil operations. That means it remains an object of great suspicion for many in the environmental movement – some of whom view Fiona and her ilk as part of a dangerous new PR exercise from the resources sector, 'greenwashing' its image to distract attention while it continues savaging the environment. Fiona has received messages asking her how she sleeps at night as an architect of this destructive charade.

Far from a publicity stunt, Fiona insists, BHP's sustainability drive is vital to ensure the company's survival in an industrial sector undergoing historic change. Like Ahmad at Aramco, she expects to see increasingly tough policies putting a price on carbon emissions, forcing companies to factor into their business decisions the true cost of the gases they pump into the atmosphere. BHP – which already forces its units to use an internal carbon price – would welcome that shift, Fiona says, since it would gain an edge over less efficient, more polluting, rivals. 'It's so obvious,' she says. 'Put a bloody price on it, seriously!'

As governments worldwide show a greater appetite for tackling climate change, heavily polluting companies have already begun to pay a penalty in the stock market. A few months after Fiona and I meet, BlackRock – the world's biggest money manager, with $7 trillion under management – sends shockwaves through markets by saying it will now put climate considerations at the centre of its investment decisions. A growing number of financial groups are going further and formally blacklisting investment in heavily polluting companies, most notably Norway's trillion-dollar national wealth fund (an ironic twist, given that the fund exists to manage the enormous profits that country has made from its offshore oil fields). In the US, fossil fuel energy firms, which once made up a quarter of the S&P 500 stock index's value, have seen that figure fall below 3 per cent, with coal companies going bust in droves as investors turn tail.

People like Fiona no longer have to plead with colleagues to cut emissions for the sake of the planet. Now it's for the sake of the share price, too. As Morrison's government works to save projects like the Adani mine, and to protect coal jobs in places like Mackay, it increasingly looks to be flailing against the onset of a historic shift – albeit one that is long overdue – in the global market economy. 'This idea of people sitting in dark smoky rooms planning how to impede action on climate change – I think that's largely gone,' Fiona says. 'There's a momentum in the system that can't be stopped.'

CHAPTER 19

If global temperature temporarily overshoots 1.5°C, Carbon Dioxide Removal would be required to reduce the atmospheric concentration of CO_2 to bring global temperature back down . . . The larger and longer an overshoot, the greater the reliance on practices that remove CO_2 from the atmosphere.

– Intergovernmental Panel on Climate Change,
Global Warming of 1.5°C. An IPCC Special Report (2018)

HELLISHEIÐI, ICELAND

About 200 million years ago, unnoticed by the dinosaurs that had begun to dominate its surface, the supercontinent Pangaea started to break into pieces. As the Americas drifted apart from Europe and Africa, a great ocean formed between them with a constantly stretching rift on its floor, through which roiling magma spewed up to create an underwater mountain range, 10,000 miles long. In an area of particularly violent activity towards the range's northern tip, the lava reached the sea surface and formed an island, its craggy terrain remodelled endlessly by the eruptions that continued with savage frequency over millions of years.

Iceland is still being slowly rent asunder as the Eurasian and North American plates continue their long divorce. At various points on the island you can stand on either side of deep ravines that lie between the plates, widening by 2 centimetres each year. The tectonic fault has given this country an extraordinary concentration of seismic activity: 130 volcanoes in an area smaller than Guatemala. They've posed a threat of chaos and destruction throughout Iceland's eleven centuries of human habitation. One eruption in the 1780s triggered a famine that killed a quarter of the population. Another in 2010 forced airlines across Europe to cancel more than a hundred thousand flights.

But in recent decades, Iceland has started to reap benefits from the magma churning beneath it. It's built one of the world's most advanced networks of geothermal power plants, where turbines are driven by steam bursting from the hot bedrock. Together with hydroelectric dams spanning rivers across the island, this means Iceland now derives 99.99 per cent of its electricity from renewable sources (the other 0.01 per cent: diesel generators used in emergencies and on a couple of remote islands). It's one of the greenest energy systems on earth. But Edda Aradóttir isn't satisfied with Iceland slashing its own carbon emissions. She wants it to start sucking up other people's.

We meet at the Hellisheiði power station, Iceland's biggest geothermal plant, which provides the capital Reykjavik with electricity as well as scalding hot water that heats the city's buildings and also its streets, to keep them clear of ice in winter. Nestled in the lee of a hulking volcanic outcrop, the plant's chimneys send languid tongues of steam into the late summer sky, suffusing it with the eggy smell of sulphur from the basalt bedrock beneath. That bedrock, Edda says, has the capacity to lock away more than 3 trillion tonnes of carbon dioxide under the Icelandic soil: double the amount that humanity has emitted since the dawn of the Industrial Revolution.

For now, making even a perceptible dent on those accumulated emissions remains a colossal challenge – as Edda knows better than anyone, having been working on it for her entire career. She had taken a summer job at Reykjavik Energy in 2006, after finishing a master's in theoretical chemistry, when the power company asked her to help it start work on Carbfix, an unusual new project spearheaded by Icelandic president Ólafur Ragnar Grímsson. The green-minded head of state had approached Wally Broecker, the legendary US scientist who had been raising the climate change alarm since the 1970s, to ask what more Iceland could do to tackle the problem. The discussion quickly focused on the extraordinary potential of Iceland's volcanic bedrock to turn huge quantities of carbon dioxide into inert stone.

The basic science behind the idea was elementary. Volcanic basalt like the stuff under Iceland is rich in calcium, which reacts with carbon dioxide to form calcium carbonate, better known as limestone. It's been happening naturally for billions of years. In principle, it seemed an obvious way of safely locking away excess carbon dioxide. But no one had established how long the mineralisation process took, and many scientists imagined it would be a matter of centuries. In that case, there would be little practical difference from the methods being explored by companies like Aramco, which simply aim to store carbon dioxide underground in gas form – prompting critics to warn that it could leak back into the atmosphere at some point.

Edda took a permanent position at Carbfix to study this question, submitting her findings in her 2011 PhD thesis. She concluded that under the process the team had designed with Broecker's lab – dissolving carbon dioxide in water and injecting that 'soda water' hundreds of metres underground – most of the carbon would be turned into limestone within just three years. She got her doctorate, but many in the wider scientific community questioned whether her theoretical work would stand up in the real world.

Edda drives us in her electric VW Golf to a small silver dome, looking like part of the set from an early episode of *Star Trek*. A few sheep hover nearby, drawn from nearby fields by the warmth of the pipes that run here from the power plant. Inside the dome is the hefty pump that sends the carbonated water underground through a 2-kilometre-long pipe. Carbfix started testing the process the year after Edda published her paper, using some of the carbon dioxide released by the geothermal power plant – tiny quantities compared with fossil fuel stations, but enough for their experimental purposes. They collected 200 tonnes of carbon dioxide, dissolved it in water and pumped it underground, tracking the mineralisation progress using tracer particles and a collection of monitoring wells dispersed across the area. Two years later, the measurements showed that 95 per cent of the carbon had already become stone. The team drilled into the bedrock to collect a sample, which Edda shows me now inside the dome: a cylinder of black rock, studded with white dots of limestone.

Science published the results under the headline 'Inject, Baby,

Inject!'. The science was now clear: effectively limitless amounts of carbon could be stored in the basalt rock that lay not just under Iceland, but beneath large chunks of every continent, and great swathes of the ocean floor. The flow of fossil gases could be shoved into reverse gear, with planet-warming carbon molecules put back where they started, underground and inert. The question now was how to get hold of the carbon.

* * *

The 2016 UN climate conference in Marrakech was a downbeat affair, overshadowed on its second day by the election of Donald Trump, with his vow to pull the United States out of the global agreement signed in Paris a year before. But the trip to Morocco was worthwhile for Ólafur Ragnar Grímsson, newly retired after twenty years as Iceland's president, who got talking there to a curly-haired young German engineer with an intriguing business.

By this point, it was seven years since Christoph Gebald had set up his company, Climeworks, with his college friend Jan Wurzbacher. The two had met on their first day at ETH Zurich, the famous Swiss institution where Albert Einstein taught physics after failing to gain entry as a student. Each had long wanted to become an entrepreneur and they quickly decided to do so together. Whatever they did, they agreed, would need to target one of the world's big problems. 'We would never have started a company to build a translation app for the iPhone or whatever,' Jan says now. Instead, during the course of their mechanical engineering studies, they started designing machines to pull carbon dioxide out of the air.

The logic behind Climeworks was grimly pragmatic. However hard the world tried to reduce its carbon emissions, Jan and Christoph had concluded, the planet would still be headed for dangerous levels of warming. Even if emissions fell by 90 per cent, the amount of carbon dioxide in the atmosphere would still keep ticking up – still pushing us towards catastrophe, only more slowly. Decades after the first attempts at international action on climate change, emissions were still rising every year; China and India continued to build coal-fired plants apace, and there was no sign of viable emissions-free technology for giant industries like aviation and shipping. The idea of a zero-carbon global economy in the foreseeable future seemed fanciful. But with technology to remove carbon from the air – direct air capture, or DAC, as it would become known – the world might still hold carbon dioxide levels steady, or even reduce them, eventually, to the concentration seen before the Industrial Revolution.

The early days of Climeworks were tough for its young founders. 'Nine out of ten people were just punching us in the face,' Christoph tells me. It seemed impossible, their doubters said, that they would be able to produce their machines at anything like the scale needed to make an impact on the climate. In fact, some charged, their work could be counterproductive. If policymakers and the public started pinning their hopes on DAC, it could cripple the already struggling effort to wean economies off fossil fuels. In 2011, just as Jan and Christoph were trying to raise their first investment round, the venerable American Physical Society published an exhaustive 100-page report on the prospects for their nascent field. Written by thirteen prominent experts from the US and Europe – including one of ETH's top professors – its verdict was withering. 'DAC is not currently an economically viable approach to mitigating climate change,' read the first of its *Key Messages*, before a warning of 'extreme caution' against any strategy that relied upon negative emissions technology.

The duo still managed to raise enough cash, principally from wealthy Swiss families, to start making their 'collectors' – steel cubes about two metres on each side, with a fan at one end and an air vent at the other. Inside each is a sort of sieve coated with a chemical substance that Jan and Christoph developed, which absorbs carbon dioxide from the air sucked in by the fan. When the filter is saturated – several times a day – the box closes and starts heating to about 100°C. The heat makes the carbon molecules detach from the filter, and the box fills with pure carbon dioxide that can now be piped away for disposal.

Climeworks is not the only startup pursuing DAC. Their most prominent rival is Carbon Engineering, a Canada-based company built on the work of Harvard professor David Keith, which has attracted hefty funding from the likes of Bill Gates and oil giant Chevron, as well as Fiona Wild's BHP. But Keith's system requires temperatures several times higher than Climeworks', meaning it must be deployed in large-scale plants and – in its current iteration – relies on burning natural gas to generate the required heat. In contrast, Climeworks' boxes can be assembled into a plant of any size, and can draw the more modest amounts of heat needed from a wide range of industrial and natural sources.

All the same, you would need a hell of a lot of those boxes to make an appreciable impact on atmospheric carbon levels. Each Climeworks collector can absorb 50 tonnes of carbon dioxide a year, roughly equivalent to the carbon footprint of a single US family. The annual emissions caused by all human activities – of carbon dioxide alone, excluding other greenhouse gases – amount to well over 30 *billion* tonnes. But if the world decided to pursue DAC on a massive scale, it would be easily possible, Christoph insists. Each box is roughly the size and weight of a family car, and less complicated to produce, he notes. And the world manages to make nearly 100 million cars a year without trouble.

When Christoph met Grímsson in Marrakech, Climeworks was about to unveil its technology to the world by launching its first plant, an array of eighteen collectors in Hinwil, a village near Zurich. Carbon dioxide from the plant would be sold to Coca-Cola's Swiss subsidiary, which would use it to put the bubbles in its fizzy drinks. Christoph and Jan were still considering what kind of approach could be suitable for their ultimate goal of locking away the carbon permanently. Grímsson decided to make another introduction.

The partnership between Climeworks and Carbfix opened a new chapter in the young history of carbon offsets: a field that has been dogged by controversy since its inception. Under one UN programme launched early this century, the Clean Development Mechanism, rich-world polluters could cancel out their emissions by funding renewable energy projects in developing countries. The biggest chunk of this money was taken by China to build wind and hydroelectric plants which, critics argued, it would have built anyway. As that scheme faded, tree-based offsets took centre stage, aimed at either planting new areas of woodland or protecting existing ones. It was often hard to gauge the impact of such schemes. Protecting one section of a rainforest from deforestation might just mean the loggers cut down the next patch instead. And the amounts of carbon capture promised by tree-based offset schemes often counted on those trees living for a century, without proof that they could be protected for that long. Amid all these efforts, the world's forests have continued to shrink at a devastating pace, losing a tenth of their area since 2000.

For trees to capture enough carbon from the air to hold back climate change, we would need to abruptly turn that shrinkage into growth at a head-spinning scale. Even if you could somehow create and preserve a new forest larger than Canada, it would cancel out less than three years of greenhouse gas emissions at our current rate. In that context, the Climeworks vision may seem less outlandish. 'Scaling up machinery is something we've proven we can do quite well, as humans,' Jan says. 'Putting up a new industry at a global scale in thirty years is a big challenge. But it's not impossible.'

In 2018, the credibility of this new industry received a massive boost with a landmark report from the IPCC, focused on the now highly ambitious goal of limiting global warming to 1.5°C. 'All pathways that limit global warming to 1.5°C with limited or no overshoot project the use of carbon dioxide removal on the order of 100–1000 gigatonnes over the 21st century,' the report read, suggesting that direct air capture had a greater maximum potential than tree planting. Having been dismissed as an unhelpful distraction, their field was now seen by the scientific establishment as a possible game-changer.

A few months later, Jan and Christoph launched a new breed of carbon offset scheme. The first collector at Hellisheiði had been installed, using geothermal heat to capture carbon that was being mineralised underground with the Carbfix system. To fund an expansion, Climeworks set up a webshop to sell rights to the carbon that would be captured by a full-scale plant. The price came in at over $1,100 per tonne, a sizeable markup over what Climeworks says is its current operational cost of about $600. Its long-term goal is to get down to something near $100. All these numbers are orders of magnitude above the price of tree-based offset certificates – often well under $10 per tonne. But within the first fifteen months of the scheme, more than four thousand people had paid for Climeworks carbon certificates – including Bill Gates, a key investor in rival Carbon Engineering, who boasted of having bought enough Climeworks offsets to get a volume discount. Gates' old company Microsoft signed up too, as did the carmaker Audi and the financial tech group Stripe.

At Hellisheiði, Edda shows me the patch of barren ground where assembly is about to begin of the next phase of the capture plant: eighty Climeworks collectors, with an annual capacity of 4,000 tonnes. She gives short shrift to complaints that this technology will distract from renewable energy development. Had the world moved more quickly to tackle the climate threat when scientists first raised the alarm, it would have been able to pick and choose its methods, she says. Now, after decades of delay, it must attack the problem with every weapon available. 'We're past the point of having the luxury to debate this,' Edda says. 'We have to take action. We absolutely have to take all the steps we can.'

* * *

'Have a gulp,' Carl Berninghausen grins as I pick up a beaker of crude oil from a table in his office, in the eastern German city of Dresden. Without the orange hazard stickers on the container, the clear liquid could pass for vodka. It's just useless dirt and impurities, Carl explains, that create the black hue of the oil pumped from the ground. This is pure e-crude – created by his company Sunfire using water, carbon dioxide and electricity. Instead of burying captured carbon underground, Carl wants to use it to make renewable fuels that could tackle some of the most treacherous challenges within a shift to a carbon-neutral economy.

A century ago in nearby Berlin, scientists Franz Fischer and Hans Tropsch found a way to make liquid hydrocarbon fuels from carbon monoxide – obtained by heating coal – and hydrogen. It was a potentially transformational development for countries lacking easy access to oil. Unfortunately for the historical legacy of Messrs Fischer and Tropsch, the most enthusiastic users of their process were two of the twentieth century's most evil regimes. Adolf Hitler used it to power his war machine with fuel derived from Westphalian coal; later it helped sustain the sanctions-hit Apartheid regime in coal-rich South Africa. Now the system might have a shot at spectacular redemption in the climate battle. There's no reason why the carbon used in the process must come from coal, Carl says. Sunfire has created an advanced system, built around the Fischer–Tropsch process, that instead uses carbon dioxide as a primary input. When used with Climeworks' technology, the system churns out petroleum fuel that emits only as much carbon as was taken from the air to produce it.

A lean man in his fifties who founded Sunfire after making a fortune in the lumber industry, Carl faces a struggle to convince some in the environmental movement that there is any long-term future for hydrocarbon fuels, even of the carbon-neutral sort. But while electric cars are already a viable alternative to the petrol-powered variety, electrifying other forms of transportation will be far harder – not least commercial aviation, which is set to boom as demand takes off in developing countries like India.

It's thanks to the extraordinary energy density of jet fuel that a crowd of several hundred people can fly over oceans inside a closed metal tube. One kilogram of jet fuel provides as much energy as 50 kilograms of lithium-ion batteries. To grasp the scale of the obstacles facing electric flight, try to imagine a Boeing 767 – which weighs some 220 tonnes with a full tank of jet fuel – taking off while carrying 3,600 tonnes of batteries. Another potential option being discussed in the industry is hydrogen power. But hydrogen-fuelled jumbo jets are many years away even in the most optimistic scenario. And if that breakthrough does come, moving aviation to hydrogen would mean replacing tens of thousands of planes worth hundreds of billions of dollars, along with all the fuel infrastructure that supports them – a transition that would take years more to complete, with atmospheric carbon levels ticking up all the while.

In contrast, Carl says, the technology to make carbon-neutral 'synfuels' is already here and can be put to work right away to power planes, ships and other forms of transportation (not to mention, he adds, fossil petroleum-based cosmetics, which currently have people putting dead animal remains on their faces). Those pushing for a wholesale abandonment of hydrocarbon fuels, he argues, will need to show greater realism and flexibility, or risk sidelining a vital tool in building a carbon-neutral economy. 'There are still people who think that carbon dioxide is bad, so fuel made from it is bad,' Carl says. 'Well, is it a better idea to just leave that carbon dioxide in the air?'

On the edge of a fjord in southern Norway, Carl's vision is beginning to take shape, through a consortium called Norsk E-Fuel, formed by companies from across Europe. A Luxembourg engineering firm will build a synfuels plant using Sunfire's technology to make 100 million litres of e-crude per year, with carbon dioxide from an array of Climeworks collectors. The plant will run on

Norwegian hydroelectric power, with its output taken by Finnish energy group Neste to be refined into fuel in the Netherlands.

It's a case study in the seamless cross-border collaboration enabled by the single market that covers – bar the United Kingdom – every country in central and western Europe, including non-European Union members like Iceland, Norway and Switzerland. And through their new European Green Deal platform, the bureaucrats of Brussels want to make the continent a leader in the technologies that will avert the worst impacts of climate change.

'Europe needs a mission,' Frans Timmermans tells me on a video call from his office in the Belgian capital. A jovial white-bearded Dutchman who trained as a prisoner of war interrogator during the Cold War, the multilingual Timmermans rose to serve as Dutch foreign minister, and then as deputy to Jean-Claude Juncker, president of the European Commission. When Juncker stepped down in 2019, Timmermans was at one point the favourite to replace him as head of the European Union's executive branch, but found his path blocked by eastern European leaders whom he had criticised for eroding civil liberties. The top job went instead to German defence minister Ursula von der Leyen. But she retained Timmermans as second-in-command, and put him in charge of a policy drive that she had put at the heart of her presidency, one that he had himself promoted during his own ill-fated campaign: making Europe a leading power in the new green economy.

For decades, Timmermans says, the driving force behind the European Union was a desire to leave behind the carnage of warfare that had ravaged the continent for centuries. As the years passed, new hazards provided further spurs for cooperation: Soviet aggression, for example, or financial crises. 'Very often Europe was looking for a sort of defensive mission,' the vice-president says. 'What I find fascinating with the Green Deal is that it's *offensive*.'

Timmermans' vision of reshaping Europe's economy faces opposition from vested interests – notably his old critics in Poland, with its heavy reliance on coal. To ease the pain for the economic losers of the green transition, he is developing the Just Transition Mechanism, an initiative that will offer billions in financial support to places that currently rely on polluting industries. Meanwhile, he's fighting to ramp up the firepower behind innovation funds like the Horizon programme that has been a major financier of the Carbfix/Climeworks collaboration.

The single greatest boost for such companies' long-term prospects could be a powerful, comprehensive carbon pricing apparatus. Europe already has an emissions 'cap and trade' system, launched in 2005, the first and – until China launched a similar scheme in 2021 – biggest scheme of its kind in the world. Specified polluting industries are issued a certain number of emissions permits. Companies that are making faster progress on cutting emissions can sell their spare permits to laggards, and pocket the profit. But the scheme only covers about half the continent's emissions: electricity generation and some heavy industry is included but the vast bulk of transport pollution, crucially, is not. Meanwhile, generous distribution of permits has kept their price languishing well below the level that leading analysts reckon is needed to drive a real economic transformation. But at the right carbon price, Sunfire's renewable jet fuel or Climeworks' offset certificates could quickly start to make cold business sense.

Expanding the permit scheme into areas like long-distance aviation and shipping, and ensuring the carbon price creates serious financial incentives, is a top priority for his team, Timmermans says. And to make sure European businesses stay competitive, they're looking into measures to levy tariffs on cheaper, high-carbon imports from countries with laxer policies. 'We're absolutely justified in protecting our industry and our society against false competition and carbon leakage,' Timmermans says. The plan has prompted pushback by trading partners from Moscow to Washington. But Europe's huge, prosperous market is one that no one will want to be left out of, he says. And by raising its standards on carbon emissions, he hopes it could galvanise a wave of similar measures around the globe.

'We don't have to defend ourselves against the world,' Timmermans says. 'No – we could lead the world, and others will follow us. It's the first time in quite some time that Europe could play this leading role.' But as they jostle for supremacy in the decades ahead, leadership in green technology will be one of the most fiercely contested prizes among the world's superpowers.

PART SEVEN

Power

CHAPTER 20

Building an ecological civilisation is vital to sustain the Chinese nation's development . . . Comrades, what we are doing today to build an ecological civilisation will benefit generations to come.

– Chinese president Xi Jinping, addressing the 19th National Congress of the Communist Party of China, 18 October 2017

SHENZHEN, CHINA

The airtight walkway runs for nearly a mile through one of the world's most advanced battery factories. Below, an array of yellow robots jerk and pirouette through the stages of production – fetching the powdered raw materials of the electrodes for mixing into liquid form, layering the black paste onto narrow metal sheets in constant motion like a four-lane motorway, baking them at a heat that warms my face even in my sealed-off isolation. Towards the top of the high walls are green banners calling for 'Passion and Innovation!', visible only to a scattering of operators, engineers and quality inspectors who come and go in colour-coded uniforms.

From this plant at its Shenzhen headquarters and four others elsewhere in China, BYD can produce 65 gigawatt-hours of batteries each year – enough to power six billion iPhones – and it's still pouring huge sums into a frantic expansion of that capacity, to position itself for a historic shift in the world's energy system.

Founded in 1995 by Wang Chuanfu, a former government metallurgical researcher in his late twenties, BYD started out making batteries for handheld electronics, getting its first big break with contracts to supply mobile phone pioneers Nokia and Motorola. Now, the world is counting on lithium-ion batteries not just to power our phones, but to play a central role in weaning the modern economy off fossil fuels. For BYD and a fast-growing cohort of Chinese peers, that means a chance to build a corporate empire for the green energy era.

The smell of paint hangs thick in the air of one of the world's biggest electric car plants, deep within the BYD headquarters complex. Yellow-helmeted workers scurry with the urgency of Formula One pit crews to fasten battery packs into the suspended chassis drifting overhead, keeping to a pace that sends a completed vehicle rolling off the end of the assembly line every ninety seconds. BYD has built more electric vehicles than any other company in the world, driving an electric-first future that is already conspicuously taking shape in its home city. Public transport on Shenzhen's roads is now entirely electric, with 16,000 BYD buses and 22,000 of its e6 taxis humming gently amid the skyscrapers and neon-decked plazas.

Snaking a 3-mile route through BYD's Shenzhen enclave, the company's newest battery-powered mode of transport is on show – an elevated electric monorail system, with the train recharged at every stop, that has already been deployed at commercial scale in the Chinese city of Yinchuan, with another large system under construction in the Brazilian port city of Salvador. Across its campus, the group has been installing rooftop solar panels – yet another major business line – which send their electricity to BYD battery storage systems for twenty-four-hour use in the buildings where 40,000 employees work and, in most cases, also sleep.

'We're the one manufacturer in the world that can offer this total green solution,' says Ken Chen, a BYD veteran who joined the company in the 1990s and now leads the solar business. 'We're not just selling products. We're selling a dream.'

Few places on earth have undergone such profound transformation in a single generation as Shenzhen, which in 1977 was a farming settlement with 20,000 residents in southern Guangdong province, remarkable for nothing except its proximity to the border with Hong Kong. In those days the biggest source of excitement here was the tens of thousands of young mainlanders who crept through the area heading south, risking everything they had in a bid to sneak across the border and seek their fortunes in the British territory's thriving market economy, where the average income was nearly twenty times higher than in China.

Briefed about the problem during a visit to the region, a

5-feet-tall septuagenarian official named Deng Xiaoping gave an early sign of the pragmatism that was to reshape his nation. The answer, he said, was not tougher border security, but to give youngsters in Guangdong better reasons to stay. After Deng became China's de facto paramount leader the following year, he followed through on his words. Under Xi Zhongxun, a hardened former communist guerrilla, Shenzhen became a petri dish for an experiment with capitalism that would shake the world.

Four decades after the birth of the Shenzhen Special Economic Zone, the town of 20,000 has become a megacity of 13.4 million, a colossal manufacturing powerhouse with annual export revenue exceeding that of Brazil. The experiment with private investment that began here was rolled out across the whole country, lifting hundreds of millions of Chinese out of poverty, with national income rising nearly fortyfold. It was a boon for corporations and consumers around the world, who rushed to take advantage of the surge in low-cost manufacturing. But for the global climate, China's economic miracle was a disaster.

No less than the Western ones that preceded it, China's economic lift-off was built on fossil fuels. In the decades after Xi Zhongxun opened the Shenzhen Special Economic Zone, China began burning more coal than the rest of the world combined, opening new coal-fired stations at a rate of about one a week. Transport emissions were also surging along with sales of cars – previously forbidden to all but senior party officials – which reached the highest volume of any country. The trendlines led to a grim conclusion: that China, now by far the world's biggest polluter, would smash the last hope of preserving a stable global climate. If that disaster is to be avoided, BYD, and the rest of China's fast-growing army of clean tech companies, will need to set this country's carbon-spewing economy on a fresh path.

* * *

From her office on the twenty-second floor of a glass Shanghai tower, Yisha He and I look down at one of her first solar installations on a nearby rooftop – hundreds of midnight-blue sheets arranged in long rows, feeding power into the shopping centre beneath. After seven years in business, her company Unisun Energy has now installed these solar arrays much further afield, from Rotterdam Airport to the plains of Adilabad in central India.

Still buzzing with energy as the sun descends over Shanghai, despite juggling startup leadership with the demands of her four-month-old son, Yisha says China's breakneck transformation during her lifetime should be seen merely as a prelude to what will come next. 'People say the last twenty years was fast – yes, it was,' says Yisha, the 33-year-old Oxford-educated daughter of a real-estate developer in the port city of Ningbo. 'But the next twenty will be even faster. Our generation is different from our parents – more and more have gone abroad to study, and just think differently. And this will be the dominant population in the economy very soon.'

Businesses like Yisha's have been on a white-knuckle ride since the rise to power of President Xi Jinping. Son of the old soldier who kickstarted Shenzhen's development, and with it China's coal-fired economic boom, Xi is charged with reinventing the smoke-belching powerhouse seeded by his father, and making China a leader in the sustainable technologies that will fuel the future. At the convention that installed him as national leader in November 2012, Xi oversaw a change to the Chinese constitution, enshrining the drive for 'ecological civilisation' – a loosely defined green agenda – as a permanent national priority. A few weeks later, the urgency of that push was thrown into stark relief when pollution in Beijing's notoriously smoggy skies spiked to record levels, with long-suffering residents voicing new levels of discontent as foreign media declared an 'airpocalypse'.

Chinese solar power was already primed for lift-off. In the years before Xi's accession, the government had nurtured the world's biggest solar manufacturing sector by bestowing billions in grants and tax breaks. At first the panel makers relied on demand from Europe and the US – until the state set up a new system under which it promised to buy electricity, at generous prices, from new solar parks. That sparked growth which, in the early years of Xi's rule, turned explosive.

Despite technological advances that were fast driving down the cost of solar panels, Xi's administration imposed only modest annual reductions on the electricity tariffs it promised to park developers, allowing their margins to head skywards. Lured by the smell of guaranteed profit, entrepreneurs flocked to set up plants and panel manufacturers ramped up their output to supply them – and as production boomed, panel prices fell even more rapidly than before, further fuelling the frenzy. Like the rest of her industry, Yisha felt the benefits, taking on a fast-growing stream of new business across China and beyond. She was also dipping a toe in the political waters, winning a seat representing Ningbo's 8.2 million inhabitants on the city's people's congress. Her foreign friends seem to view China's one-party rule as oppressive, she reflects. 'But it's difficult to manage a population of 1.4 billion people. And the good thing is, when we want to do something, we do it fast.'

Too fast, it seemed to some. During Xi's first five years in power, China built more new photovoltaic solar capacity than the cumulative total of all the plants ever built in the world before his arrival. Most of this was built in remote locations across China's picturesque far west, endowed with bountiful land and a clear, sunny environment but (by Chinese standards) few people, thousands of miles from the country's major population centres. Huge amounts of the electricity generated went to waste, and the government agency responsible for buying it fell into a $15 billion deficit.

So, five years into Xi's ecological civilisation mission, his government hit the brakes on its solar drive, suddenly declaring that no new solar farms would be eligible for financial support. Investment fell sharply, sparking fears for the survival of weaker companies in the sector. Some commentators wondered whether Xi's China was sidelining its green agenda to reinforce its focus on an economic growth rate that had been slowing. Yisha has a more bullish perspective. The government's financial firehose was switched off, she insists, because it had achieved its purpose. The investment boom had enabled Chinese companies to corner the market in solar panels, with two-thirds of global production. In the process the leading companies have achieved huge efficiency gains, sending the cost of solar power crashing more than 80 per cent in the past decade.

China's own pace of installation may have slowed from its frenzied peak, but it's still building more solar farms than any other country – and the rest of the world is now adding solar capacity more rapidly than ever before, thanks to the cost gains achieved by Chinese factories. Solar power is now cheaper, in many parts of the world, than fossil-fired plants. It's the historic tipping point that renewable energy cheerleaders had been predicting for years, reached far earlier than most of them had imagined possible.

Yisha is now gearing up for a new chapter in the solar story, as simple economic logic pushes businesses and households to harvest electricity cheaply from their rooftops rather than rely on costlier, dirtier grid power. Her window view of solar panels in dark rows will soon become near-ubiquitous, she predicts, across Shanghai and other cities throughout the world. It's the right time, she goes on, to give a fresh jolt of market forces to her sector, to weed out the weak and push the strong to develop faster. 'Those companies that don't have an advantage or a core value, they . . . they'll die,' she says. 'And I think it's good for this industry to kind of kill some of the companies and get new blood in.'

* * *

The next day, Chinese media is abuzz with scenes from Giga Shanghai, Tesla's first overseas car factory, which has been thrown up in less than a year in the city's industrial southern district. On a stage before hundreds of whooping fans, the company's mercurial chief Elon Musk tosses his jacket aside to perform an awkward dance before presiding over a handover ceremony to congratulate the plant's first customers. It's an ominous moment for the likes of He Xiaopeng – like Musk, a wealthy web tycoon turned carmaker, and now one of his most prominent rivals in the world's biggest electric vehicle market.

I meet Xiaopeng after taking a test drive of one of the first units of his newest car, weeks before its public launch, at his group's headquarters in the southern merchant city of Guangzhou. Xpeng Motors' P7 is its fastest model to date, its acceleration shoving my torso into the driver's seat with an engine that never rises above the gentle buzz of an upmarket washing machine. Depending on the configuration, the P7 can reach 100 kilometres per hour in 4.3 seconds – faster than a BMW 7 Series – or travel more than 700 kilometres between charges, a far longer range than most electric cars.

Quietly spoken and soft-featured, with cartoon figurines amid the photographs and trophies behind his desk, Xiaopeng is one of more than a thousand Chinese people to have become dollar billionaires amid the furious growth that began with the 1980s reforms. Like many younger members of the club, he was the creator of a tech company, UCWeb, which offered a lightweight mobile browser that became the most popular in China and other Asian markets. After it was acquired by e-commerce giant Alibaba in China's biggest ever internet buyout, Xiaopeng found himself with a senior job in the country's most valuable company – and with a large fortune to invest.

His interest was soon grabbed by two young engineers who approached him with an idea for an electric car company that would emulate Tesla's slick looks and high performance, but at a lower price. After his hefty early investment, they renamed the company after him – partly out of gratitude, but also because they thought the name sounded like a winner, its second syllable evoking a giant bird from ancient mythology.

Like Musk at Tesla, Xiaopeng moved from bankrolling the car company into becoming its full-time leader – a decision he made in 2017, soon after the birth of his second son, which set him thinking about the legacy he would leave to his children. 'A good entrepreneur has to provide value to the world,' he says. 'It's not just about money.' He was walking out on a Chinese e-commerce market that had become a money-spinner like few others in history – in one day of that year, Alibaba enjoyed sales of $25 billion. But the electric car market was just hitting its ignition point.

No less than solar power, electric cars were at the centre of Xi Jinping's ecological civilisation agenda, with its twin goals of cleaning China's air and making it a new energy superpower. Chinese car buyers went electric in their hundreds of thousands, drawn by government subsidies worth thousands of dollars on every sale – as well as the added incentive of skipping the tedious lottery system that limited purchases of traditional cars in the big cities. Xpeng rode the wave, with strong sales of its first car, a sleek sports utility vehicle – even as its boss broke industry ranks to call for the subsidies to be dropped. 'There were a large number of new companies making very bad, very cheap vehicles in order to get subsidies,' Xiaopeng says. 'I think this is of no value.'

To the dismay of most of Xiaopeng's peers, a few months before our meeting China had slashed subsidies for electric cars by half, sending sales plummeting. Xpeng's main rival Nio – which had revelled in a high-profile New York flotation the previous year – warned investors that it was fast running out of cash, before being rescued by a bailout from the government of Hefei, a central Chinese city, with tough conditions attached.

But for Xiaopeng and his financial director Brian Gu – a high-powered banker lured from a role as JPMorgan's top man in

Asia – this is all part of a shakeout that will enable the strongest companies to forge ahead while weaker rivals fade into irrelevance, as seen in every part of the hard-charging technology sector. Tesla's assault on China, they argue, will provide a fresh boost to the country's entire electric vehicle market – and as long as Xpeng can maintain a grip within it, keeping quality high and costs low, it will benefit. With technological progress continuing to drive down battery costs, the car market is approaching the kind of transformational moment seen a decade earlier with the sudden rapid adoption of smartphones, Brian claims. 'It's like an explosion. All those cars on the street are going to be replaced by some form of new energy. The disruption is certain.'

* * *

Protruding from atop a white pole as tall as the Great Pyramid at Giza, each the length of two blue whales, the blades of one of Lei Zhang's wind turbines spin in the thin winter breeze over the eastern city of Jiangyin. Every day this rotor, its three white arms curved like the limbs of an insectoid robot, generates 19 megawatt-hours of electricity – enough to power several thousand Chinese homes and double the requirements of the Jiangyin campus of Lei's Envision Group. Behind the bullet nose at its centre is a thicket of sensors and lasers, enabling the turbine's software to anticipate shifts in the speed and direction of the wind, and twist the blades to maximise power delivery and avoid damage. 'It's like tai chi,' Lei smiles.

Lei has needed to show nimbleness himself in a career that has taken him from the engine room of global finance to an audience with the Pope. He was in London in the years leading up to the global financial crisis of 2008, working on energy trading for the British bank Barclays and then for Total, the French oil giant. Operating deep in the guts of the modern industrial economy, Lei became uneasy about its damaging fallout. 'A capitalist looks at the

P&L for himself and his company,' Lei says, referring to the profit and loss page that is the key section of a financial report. 'But who's looking after the P&L of humanity?'

The year before the financial crash, Lei returned to China and set up a company to make high-tech, advanced wind turbines. If the technology kept improving, he'd realised, wind could soon be cheap enough to compete on cost with fossil fuel power. And China's well-resourced universities were now turning out thousands of world-class engineering graduates – providing a technical skills base to challenge the Western groups who had pioneered wind power decades before. If China was to keep growing, Lei had decided, it needed to start outperforming the West on technology, not just undercut it on labour cost.

The innovations of Lei's engineers – some enabling Envision turbines to generate power in the faintest winds, others allowing them to survive typhoons – attracted customers around the world, putting it among the top four global wind turbine producers, alongside the industry's long-established trailblazers like Denmark's Vestas. When Pope Francis summoned the bosses of the world's biggest oil companies and asset managers to a Vatican summit in 2019, to chide them that 'energy use must not destroy civilisation', he also invited Lei. Perhaps uniquely among the assembled company, Lei remarks drily, 'I wasn't looking for redemption. I was being shown as an example. To wake them up.'

From its base in wind power, Lei is turning Envision into a sprawling enterprise covering the most strategically important new energy bases. In partnership with his old employer Total, he's moving into solar power, challenging incumbents like Yisha with a venture to install rooftop solar plants for businesses in China. He's made a move on the electric car sector through the acquisition of Nissan's battery division, and has signed a deal to provide the batteries for a new generation of electric cars from Britain's Jaguar Land Rover. Connecting the elements is EnOS, an operating system to manage renewable energy networks that is already used to control over 100 gigawatts of power generation – more than the entire capacity of the UK.

In his thin-rimmed glasses and black turtleneck, Lei channels some of the messianic grandiosity of Steve Jobs as he paints a future in which EnOS will enable smart renewable energy systems throughout the world, with households and businesses generating, storing and trading electricity. It's a vision that dovetails with one of President Xi's most ambitious ideas: the Global Energy Interconnection plan, under which China's world-leading ultra-high-voltage transmission technology – which can carry electricity with little lost energy over thousands of miles – would connect a web of renewable power plants all over the world.

There's an obvious vulnerability in Xi's vision, however, and perhaps in Lei's, too. As its central place in the evolving global economy becomes increasingly obvious, clean energy risks being caught up in a new cold war between Beijing and Washington. China's technology sector has already become powerful enough to trigger a sustained national security alert in the US, which has barred Shenzhen-based telecom giant Huawei from investment in its infrastructure, and persuaded allies to do the same. It's easy to anticipate pushback against a Chinese company – let alone the Chinese state, as under Xi's plan – holding a powerful position in Western electricity systems. Meanwhile, the country's solar exporters face suspicions of forced labour at factories in the northwestern province of Xinjiang, where China's detention of huge numbers of minority citizens has become an explosive diplomatic flashpoint.

But with his country holding pole position in so many crucial green sectors – from wind to solar, from electric vehicles to long-distance transmission – Lei seems untroubled by fears that Western resistance will hinder China's emergence as the leading power of the clean energy era. 'People tend to underestimate the Chinese contribution to the world,' Lei says. 'Probably China will save the planet.'

His bullish forecast is about to be challenged in ways that neither of us can yet imagine. As we speak, a day's drive west in the ancient city of Wuhan, dozens of people are falling ill with a mysterious new virus that will soon test the most confident predictions for the global economy.

CHAPTER 21

We can, and we will, deal with climate change. It's not only a crisis, it's an enormous opportunity. An opportunity for America to lead the world in clean energy . . .

– Joe Biden, accepting the presidential nomination at the Democratic National Convention, 20 August 2020

TEXAS, UNITED STATES

Howard Schmidt lets out a cowboy whoop as a geyser of fluid shoots up from a chamber hundreds of metres underground, gushing far into the Texas sky and spattering him with tiny droplets caught in the breeze. In the bowels of the ground beneath us lies one of the bountiful shale deposits that have kept the Lone Star State at the heart of the American hydrocarbon industry. But the liquid spraying Howard's face is water, not oil. And while most of the shale wells in Texas are aimed at feeding the world's fossil fuel addiction, Howard wants his to play a part in ending it.

In the early years of this century Texas, home to oil wells since 1901, became the centre of the 'fracking' revolution: a new system using pressurised fluids to crack open underground formations of hydrocarbon-rich shale rock, and harvest the oil and gas released. As drilling crews went into a fever of activity across Texas and many other US states, the country's oil output rose dramatically, making it the world's leading producer. But the complex process meant the costs for US shale producers would always be far higher than those of rivals in places like Saudi Arabia – an issue that became critical when Covid-19 hit.

As the scale of the looming pandemic became clear, oil traders rushed to sell ahead of the inevitable global recession, while Prince Abdulaziz worsened the pain for weaker rivals by ramping up Saudi production. US oil prices briefly turned negative as producers wondered how to store all the unwanted crude. Even when that absurd situation ended, the price remained far below the break-even point for many shale businesses. The outlook darkened further when international energy giants slashed their long-term forecasts for global oil demand, as governments promised to put massive green energy investments at the centre of their recovery strategies.

The pandemic has accelerated the shift towards a world driven by clean energy – and many would see no role in that world for the Texas drillers and pumpers who drove the shale boom. But Howard begs to differ.

We meet at dawn on a cold December morning in southwest Texas, amid flat cropland that stretches to the horizon in all directions. With us is a team of burly men in hard hats with decades of experience in pumping fluid underground to extract oil and gas. Today, however, they're using their heavy pumping equipment to chase not hydrocarbons, but Howard's vision of a subterranean storage system that could speed the world's energy transition.

A gravel-voiced Texan, Howard studied engineering at Houston's Rice University and remained there as an academic, developing cutting-edge technologies for use in the oil sector. Turning his attention to fracking, he had a sudden realisation. When frackers pumped their pressurised fluid down hundreds of metres into underground cracks, they were very slightly raising the surface of the earth above. Even solid rock has a certain capacity to flex, and the fracking was inadvertently turning the top part of the earth's crust into a giant spring. 'And you can store a whole shitload of energy,' Howard realised, 'just by flexing that rock.'

And without advances in storage, the transition to an electricity sector built on renewable power looks very challenging indeed. Solar panels are manifestly not twenty-four-hour sources of energy, and wind turbines in all but the most turbulent places often stand idle. The obvious answer is to store excess power during sunny and windy periods and release it again when generation falls. But how? Lithium-ion batteries still look prohibitively expensive for use at the scale required. 'Pumped hydro' – driving water up a mountain into a reservoir, then generating power from a hydroelectric turbine when it flows back down – involves massive engineering works even if you can find a suitable mountain.

In contrast, Howard's concept – using a variation of the process that's been used to make thousands of fracking wells in Texas – could be deployed virtually anywhere. Renewable electricity would power a motor to pump water at high pressure into a crack, or lens, deep underground. Then, when electricity was needed, the water would be allowed to flow back up, driving a small turbine in the process. 'I worked on complicated shit my whole career,' Howard says. 'Lifting a kilometre of rock is pretty stone simple. You don't even have to dig it up.'

For years Howard's idea attracted no interest. He took a job in Saudi Arabia with Aramco, and used his vacations in Texas to pursue basic tests at his own expense. At last, after he'd proved the concept in 2015, he received some investment and founded a company – named Quidnet, after a Nantucket whaling village. The company's growth accelerated in 2018, with a multimillion-dollar injection from Breakthrough Energy Ventures, a fund launched by Bill Gates, backed by some of the biggest names in global business.

Howard and I watch the team carry out the process he designed: pumping first a fluid containing minute pieces of clay, to plug tiny holes in the chamber; then water to stretch the crack and test the storage capacity. This project, carried out over months, is one of several demonstrations Quidnet is doing for regional energy providers who, they hope, will pay for full-scale systems. Before long, Howard wants to start rolling out what he calls gigalenses: huge underground water chambers each capable of storing enough energy to power thousands of homes.

And the prospects for ambitious clean-power projects look set to get a lot brighter. As well as kneecapping the shale oil business, the coronavirus pandemic torpedoed the re-election chances of Donald Trump, who had looked likely to stay in office on the strength of a well-performing economy. I've arrived in Texas in the wake of a bitterly contested campaign whose aftermath has proved still more toxic. Of all the areas of division highlighted by the election, none was more extreme than energy and climate policy. Under Trump, the US became the only country to pull out of the Paris Agreement on climate action; Joe Biden promised to rejoin it on his first day in office. While Trump made public fun of wind farms and waxed lyrical about 'beautiful clean coal', Biden pledged a $2 trillion infrastructure drive that would speed the shift to a low-carbon economy.

A fiscal conservative – like many in Texas, which Trump won comfortably – Howard is uneasy about the implications of the Biden plan for the national finances, even as it looks set to supercharge the growth of companies like Quidnet. 'It could be the best thing that ever happened to us, right?' he says. 'It's gonna be indiscriminate, a tsunami of money. The problem is, that wave will eventually crash into a shore somewhere.'

But amid the political upheaval in the US, a tidal wave of cash into its new generation of clean tech startups is already well under way – driven primarily not by government policy, but by the profit-hungry animal spirits of American capitalism.

* * *

A small crowd started to gather around the Tesla Model S almost as soon as it veered off the highway on the northern fringe of Miami, shooting across three lanes before smashing into a palm tree. From the driver's seat, 48-year-old anaesthesiologist Omar Awan, unharmed by the initial impact, stared out helplessly as the car's market-leading battery pack ignited in front of him. According to a lawsuit later filed by Awan's widow, a police officer grappled in vain with the car's futuristic door handle system, then tried to smash its reinforced windows, before being forced back by the fire that was now consuming Awan, a father of five. For hours after firefighters extinguished the blaze and extracted Awan's charred body, the battery kept bursting back into flames.

Had one of Tim Holmes' batteries been powering the car, Awan might have lived. Like all lithium-ion batteries in commercial use today, Tesla batteries are swimming in a liquid electrolyte, which enables lithium atoms to move back and forth between the cell's positive and negative sides. For decades, scientists have known that a 'solid-state' battery, a simpler design dispensing with the liquid electrolyte, should in principle be superior on every metric that matters: higher performance, longer life, faster charging, smaller, cheaper – and with greatly reduced fire risk.

For decades, this elegant theory fell apart in practice, as a long series of solid-state designs proved unworkable. Now, Tim believes he and his team have finally cracked what could be one of the most lucrative engineering puzzles of the twenty-first century. And he seems to have convinced the stock market. The day before we meet at the San Jose base of his company QuantumScape, its soaring share price made him Silicon Valley's newest startup billionaire.

An understated character with a sensible haircut and black-rimmed glasses, Tim shows no after-effects of any raucous celebrations when he greets me early on a Thursday morning at the entrance of QuantumScape's two-storey headquarters, a few blocks from the hive of private jets at San Jose's international airport. The Valley's software tycoons are facing a growing torrent of criticism, accusing them of deploying vast financial and intellectual resources to profit from degrading attention spans and political discourse all over the world. But a growing number of California's most exciting startups are now shunning social media and ecommerce to focus on technologies to fight climate change.

It's less than two weeks since QuantumScape's public flotation on the New York Stock Exchange, and the explosion in its share price has left market commentators spluttering at investors' rabid appetite for a slice of the emerging clean energy economy. Within a couple more weeks it will hit a market valuation of $48 billion, significantly higher than that of the 117-year-old Ford Motor Company.

Still, to Tim it feels like anything but an overnight success. It's about fifteen years since he began exploring solid-state battery technology as a PhD student at Stanford University, the mothership of California's tech industry; ten years since he set up the company with Fritz Prinz, his old professor at Stanford, and Jagdeep Singh, an experienced tech entrepreneur. A year after launching the business, Tim realised that the substance on which the entire development plan was based – tiny 'quantum dots' of lead sulphide – was a dead end. The team considered shutting down and giving the investors their money back, but instead persuaded them to back a new approach using a different chemistry.

The holy grail for Tim, and for everyone who had failed at this challenge before him, was the solid separator: the substance that would separate the positive side of the battery from the negative side and allow lithium atoms to pass between them, without any need for the flammable liquid electrolyte that blew up Omar Awan's car. The undoing of previous efforts in this field had been dendrites: microscopic, coral-like growths of lithium that reached from one side of the battery and penetrated the separator to form a fiery short circuit. The phenomenon terrorised QuantumScape's early efforts too, to the point that employees started attending its yearly Halloween party in dendrite costumes. But then, in 2015, came the company's defining breakthrough: the discovery of a material that conducted strong flows of lithium ions while preventing dendrites, promising a far higher level of performance than any solid-state battery chemistry yet discovered.

QuantumScape won't tell anyone what this substance actually is. Its secrecy has prompted a short-selling attack by a hedge fund claiming – to firm denials by the company – that its technical claims are overhyped. But Tim seems delighted to show me how the stuff is put together. Donning hair nets and lab coats, we pass into a room with moisture levels two thousand times lower than those outside, mildly desiccating my mouth and throat. It smells, unexpectedly, like an ice cream parlour, a vanilla scent wafting over from the machines where Tim's treasured separator is being painted onto long yellow strips that are pressed between heavy rollers to vanishing thinness. In a room off to one side, an analyst is staring at electron microscope images of the product, looking for flaws as little as 10 nanometres wide: one ten-thousandth the width of a human hair.

The completed cells, each about the dimensions of a playing card, can be stacked into 'decks' containing much more energy, for a given size and weight, than any battery on the market today, according to a stream of research being pumped out by Tim's analysts. The completed automotive battery pack should charge from 0 to 80 per cent – the industry benchmark – in fifteen minutes, against thirty minutes for a top-of-the-range Tesla and several hours for many electric cars. According to another QuantumScape study, its batteries will enable a high-performance Porsche to sustain race-level speeds for five times longer than those currently used in the Formula E competition.

Tim's team still have a lot to prove, never having put a single battery in a single car. But they've received a hefty vote of confidence from Volkswagen, the German behemoth that wrestles with Toyota for bragging rights as the world's biggest carmaker. VW is now QuantumScape's largest shareholder, and plans to have a factory producing its electric car batteries from 2024. Like all other carmakers, VW's shift to electric holds the key to its long-term survival, as a growing number of countries prepare to ban the sale of petrol-powered cars. If the VW rollout goes to plan, the QuantumScape team admit, they're not sure how they'll deal with the likely explosion of demand from everyone else in the global automotive industry.

While a foreign carmaker will be the first to install QuantumScape's technology, its story highlights how the US startup ecosystem could be one of the planet's most potent weapons in the fight against climate change. The world's best – and best-resourced – universities, amassing heaving clusters of innovative talent. The huge synergies with the world-beating industries already in existence – QuantumScape's nano-scale design, production and analysis systems are all built on techniques pioneered in the semiconductor industry spawned by Silicon Valley. Then there's the money: the battle-scarred veterans of California's venture capital industry, willing to make big bets on unproven technologies and inexperienced entrepreneurs, and to stick with them through times of crisis. 'I don't think,' Tim says, 'this could have happened anywhere else in the world, right?'

* * *

For Silicon Valley veteran Vinod Khosla, QuantumScape's explosive stock market debut came as a long-awaited vindication. Born into a military family in India, he went on to study at Stanford and partnered with two classmates to found Sun Microsystems, an IT business that would become – briefly – one of the most valuable companies in the world.

Early this century, Vinod used a chunk of his fortune to seed Khosla Ventures, a fund that would back companies tackling the world's most urgent problems – chief among these, finding the technologies that could meet the world's rising thirst for energy without accelerating its drift towards climate disaster. He spent much of the following decade and a half weathering snarky criticism as a long series of his new energy protégés bit the dust. There was KiOR, a Mississippi outfit making biofuels from woodchips, which recruited former British prime minister Tony Blair as an adviser before collapsing into bankruptcy. Nordic WindPower, an offshoot from a Swedish government project, went nowhere with its wind turbines boasting two blades instead of the normal three. Then there was the doomed Infinia, creator of the PowerDish, which reflected sunlight from a huge, mirrored dish to power a helium-filled piston.

'I know I've been wrong often,' Vinod tells me from his home in Palo Alto, his crown of porcelain-white hair resplendent even on a Zoom feed. 'But I operate on the principle that when I'm wrong it doesn't matter. If I'm successful then it really matters. That's the asymmetry.'

This, of course, is the founding creed of the Valley's legendary venture capital industry: if nine out of ten companies in your portfolio fail miserably, that's fine, provided the other one is a massive success that returns the initial investment dozens of times over. Even by the standards of his tribe, though, Vinod stands out for his defiant contrariness, publicly mocking his rivals' attachment to golf and yachting, a man so stubborn that he went to the US Supreme Court in a dispute over a public trail through a California beach property that he owns but almost never visits.

For years, his wave of investments into clean energy were knocked by sceptics as Quixotic follies – though he denies ever having been pushed to question his own judgement. 'It didn't faze me that others didn't believe the story,' he says. 'I tend to have a fair amount of confidence in my views – I would almost say arrogance.'

At last, Vinod's patience seems to be paying off. In QuantumScape alone, the value of his fund's stake has surpassed $2 billion, with several of his other energy bets – as well as Pat Brown's Impossible Foods – now fast approaching critical momentum. Clean energy, meanwhile, has gone from being an eccentric sideshow in the investment world to its hottest territory. When Bill Gates decided to make his mark on the space, he asked Vinod to help him set up Breakthrough Energy Ventures, a clean tech venture fund where the two men were joined as investors by a virtual *Who's Who* of the global business elite: from Jeff Bezos of Amazon to Jack Ma of Alibaba; from Japanese tycoon Masayoshi Son to Aliko Dangote, the richest man in Africa. The potentates of global business are waking up to the scale of the business opportunity that Vinod has been chasing for the best part of two decades.

'If you look at what we need to do to solve climate change,' Vinod says, 'there's about a dozen major areas – and each of them could be tackled by one entrepreneur. It won't be done by institutions or large ventures or talking heads from the World Economic Forum.' The stage is set, he believes, for a new generation of hyper-innovative companies that will power the next chapter of global growth – each of them 'a $100 billion opportunity'. The only question is which of the new breed of energy startups will become global titans – and who will make untold billions from their success.

'It's not just about doing the ethical thing,' Vinod says with his customary bluntness. 'This is the biggest business opportunity there is.'

* * *

Two hours' drive southeast of Atlanta, carved out of dark pine woodland, I find the remnants of one of Vinod's least successful bets. Range Fuels went bust in 2012 after receiving US government support worth hundreds of millions of dollars for its technology to make ethanol from wood. Eight years later its defunct plant is still standing, a maze of gas pipes and water towers pushing into the Georgia sky. And in the shadows of this monument to expensive failure, another startup is trying to prove that it can build a very different future for itself – and for a whole section of the global economy.

Today, the old Range office is occupied by the engineers of LanzaTech, a startup that makes biofuels not from plants but from garbage and industrial waste. For all the instinctive aesthetic appeal of making fuel from beautiful green plants, the cellulosic biofuel industry comes with some alarming downsides: food supplies squeezed by diverted crop production, and a wave of deforestation in places like Indonesia as jungle is cleared for biofuel profit. Using some of Range's old infrastructure, LanzaTech have set up reactors in which they're perfecting their microbial fermentation process to turn waste into fuel. Instead of threatening rainforests and food security, their technology could absorb vast quantities of the pollution that scars and poisons our air, land and oceans.

And the startup has now started building serious momentum. Its technology is already being put to commercial use at a Chinese steel mill, and is being rolled out in India by the giant Indian Oil Corporation. In a field next to the old Range facility, LanzaTech engineers are preparing to build the company's first jet fuel plant, well placed to service airliners flying out of Atlanta's Hartsfield-Jackson International, the world's busiest airport. 'I'm serious that we could provide the lion's share of the world's liquid fuel,' says LanzaTech co-founder Sean Simpson, an Englishman with shoulder-length hair and a constantly bubbling chuckle.

It's quite a statement for a man who started out testing his industrial process in a $30 chicken rotisserie bought from a barbecue warehouse. Born to itinerant English parents in Zambia, Sean earned a plant biochemistry doctorate in the UK and did advanced research in Japan before going to work for a biotech company in New Zealand. Tasked with finding ways to help lumber companies make more money, Sean tried to develop a process to make biodiesel from trees. More swiftly than the founders of Range Fuels, he concluded that the idea was hopeless. 'Trees, it turns out, are pretty bloody expensive,' he says. Sean and his colleague Richard Foster began talking about other options, concluding that the ideal input would need to be cheap, abundant, concentrated in specific places and posing no threat to food supplies. They were talking, they realised, about waste.

Both men were familiar with the science of gas fermentation. If they could find the right microbe, Sean and Richard reasoned, they should be able to create ethanol from waste gas – taken either directly from an industrial source, or from a high-temperature plant that gasifies municipal garbage. Scouring the academic literature, they zeroed in on a bacterium found in rabbit guts that seemed to have precisely the properties they were seeking. Using the chicken rotisserie, the two men – by now both made redundant by their ailing employer – started cultivating the microbe, which achieved all they asked of it. It was around this time that Sean contacted Vinod, sending him a single-page summary of the concept. The presentation was so slapdash that Vinod still lambasts Sean about it. But he was sufficiently intrigued to write a cheque for several million dollars, enough money to fund the pilot project that would pave the way for a commercial launch. 'Everybody agreed that we could probably make this work in five years,' Sean says. 'We were just completely unrealistic. We had no idea what it would take.'

Still, over the following decade, even as countless other clean tech startups collapsed, Sean and his team kept slogging away, moving their technology closer to the efficiency needed to take on the market at scale. In 2014 – soon after Richard died of cancer – the company moved to the US, setting up a lab in Chicago and taking over the Range site in Georgia. Things began to pick up pace. Richard Branson's Virgin Atlantic flew a plane from Orlando to London Gatwick with about 5 per cent of its fuel provided by LanzaTech – all the company had available at the time – proving that it met the required standards. At a steel plant in Hebei province, China's Shougang started churning out tens of millions of litres of ethanol a year, produced from its waste gas by LanzaTech's rabbit microbes.

While Sean works on perfecting the process, LanzaTech CEO Jennifer Holmgren, a veteran of the US petrochemical industry, is looking for ways to get it out into the world. One of the most promising is being pursued in the northeast of Japan's Honshu island where ordinary, jumbled household trash is being gasified and turned into ethanol. For all the good intentions of households who dutifully sort their rubbish into colour-coded bins, conventional recycling is a complex and expensive business – so much so that large quantities of waste marked for recycling have ended up being burned or dumped as landfill. By radically simplifying the process, Jennifer says, LanzaTech's system offers a path to a truly circular economy. And the potential for recycled waste to displace oil products goes far beyond jet fuel. Migros, Switzerland's biggest retail company, is already selling a household cleaner made with LanzaTech ethanol. L'Oréal, the world's biggest cosmetics group, is using it to make plastic packaging, as it seeks to appeal to increasingly eco-minded consumers.

As interest grows in this technology, would-be partners are lining up around the globe – as are rival sustainable fuel producers like Carl Berninghausen's Sunfire. Even so, Sean says, if you're an unknown scientist, seeking a base to turn a revolutionary idea into reality and for deep-pocketed investors to bankroll it, the US remains unrivalled. 'There's not an expectation here that everything is nailed down from the beginning,' he says, with the zeal of an immigrant living a supercharged twenty-first-century version of the American Dream. 'No one is asking you how much money you're gonna make them by when. People just want to know how great your idea is, and they know that if it gets really big, money will come. It's a little bit Wild West. But it does lead to these great, innovative companies.'

* * *

In 1985, Ronald Reagan flew to Geneva for a first meeting with Mikhail Gorbachev, the newly appointed leader of the Soviet Union. Just two years before, the superpowers had come terrifyingly close to nuclear war, when Soviet intelligence had mistaken a major US training exercise for an actual attack. By now, any war would involve the use of bombs hundreds of times more powerful than those dropped on Japan in 1945, which drew their force from nuclear fission: atoms of uranium breaking apart in a chain reaction. The new generation of thermonuclear weapons drew their appalling force from nuclear fusion: hydrogen atoms smashing together and fusing, releasing titanic amounts of energy. It is the process that powers the sun.

Now, the world's two most powerful men proclaimed, they would harness the force of nuclear fusion, not for mass annihilation, but to provide the world with an energy source fit for the approaching twenty-first century. Through an international body whose membership would swell to thirty-five nations, they began preparations for an experimental fusion reactor that would provide a blueprint for remaking the world's electricity system. Fusion plants would emit no carbon dioxide or other pollutants and would require no scarce inputs. They would produce tiny and short-lasting amounts of radioactive waste compared with the fission plants already in widespread use, and carried no risk of the kind of runaway meltdown that happened at Chernobyl a few months after the Geneva summit. And the amount of power that could be generated from the technology was, in principle, limitless. Plans were drawn up for a huge reactor in the south of France, the priciest scientific experiment in history with total construction costs estimated by the US government at $65 billion. Today, half a lifetime after Reagan met Gorbachev, the International Thermonuclear Experimental Reactor (ITER) is still under construction, with a target of generating energy for the first time in 2035.

Inside an unattractive academic building at the Massachusetts Institute of Technology, Bob Mumgaard wants to tell me why his two-year-old startup will get there first. But first, he wants to explain how fusion power actually works.

With the flip of a switch, the horizontal vacuum tube on a desk in front of us gives birth to a glowing cylinder of an otherworldly pink substance, not quite a gas nor quite a liquid, hovering between the glass edges of its container. This is plasma, the fourth state of matter, the stuff that stars and lightning are made of – and the stuff that will, Bob believes, provide humanity with clean, abundant power for centuries to come.

I put my hand on the glass and feel the heat leaking through the imperfect demonstration vacuum from the plasma within. 'Everyone's like, "Oh, it's so hot, it must be like lava,"' Bob says. 'But it's not like lava, melting stuff. It's the opposite problem. As soon as it touches something that's not hot, the plasma cools down and the reaction stops.' He looks down at the ethereal pink blob and sighs. 'You know, it basically wants to be in outer space.'

When scientists try to explain fusion power to laymen, they often start with the notion of a miniature sun. It's not a bad analogy, Bob says, but it actually understates the engineering challenge. 'The stars are bad fusion reactors,' he says. Nestled in the unfathomably vast vacuum of space, the sun can sprawl with abandon, facing no danger of touching any object substantial enough to derail its self-feeding fusion reaction. Considering the sun's size – 1.3 million times that of the earth – the amount of energy it generates is shockingly unimpressive. Per kilogram, a quietly rotting pile of compost generates much more heat. 'You have to do it way better if you want to make it work on earth,' Bob says.

We pass deeper into the building, into a tall room that houses what looks something like the control pod of an early spacecraft. It is the Alcator C-Mod, built by Bob's predecessors at MIT, and used for experiments between 1991 and 2016. When it was switched on, it contained the hottest point in the whole solar system, a plasma that reached 100,000,000°C – six times hotter than the sun's core. The machine used such enormous amounts of power that it could be activated only for a few seconds at a time, by prior arrangement with the city electric authority.

This is the existential problem that has dogged fusion power for decades. Scientists have been able to create fusion reactions since the early 1950s, when US researchers developed hydrogen bombs that used a fission blast to detonate a much more powerful fusion reaction. But municipal power plants, to state the obvious, need to operate without massive nuclear explosions. And while various teams around the world have managed to create peaceful fusion reactions in a scientific setting, these always consumed far more energy than they generated.

The key to the solution, scientists realised early on, was magnets – magnets of superlative power that could hold a mass of roiling plasma, at extraordinary temperature and density, suspended in a vacuum, preventing contact with any reaction-killing matter. ITER's reactor in France will use 10,000 tonnes of magnets, in a plant covering the area of two thousand tennis courts.

The fusion team at MIT had been conducting much of their research as part of the ITER effort. But when Bob's team discovered a far more powerful new magnetic material – opening the prospect of building fusion reactors at a much more manageable scale – they decided to pursue the concept on a separate track. ITER's vast reactor was already mostly built, after all, even if it was still years away from completion. And after all the frustrations with the vast international bureaucracy running ITER, here was a chance to see whether a startup could move more nimbly.

Nimbly, in the context of fusion power, doesn't mean making money any time soon. But Bob found no shortage of investors, together willing to bet almost $200 million on his new company, Commonwealth Fusion Systems. Vinod's fund invested a large sum, as did Breakthrough Energy and several other US funds. Millions more came from Singapore's main sovereign wealth fund and the European energy companies Eni and Equinor, both anxious to expand beyond their traditional oil businesses.

We stand on a raised viewing platform, watching a team of Commonwealth employees install power cables for a forthcoming test of Bob's superconducting magnet, which will be 400,000 times stronger than the earth's magnetic field, and about double the strength of ITER's magnets. Within a few months, at a site somewhere in the northeastern US, Bob's team will start building SPARC – which, according to their calculations, will be the first fusion machine in history to generate more power than it consumes. After SPARC will come ARC, a commercial-sized plant that could provide a model for reactors to be rolled out in their thousands around the world.

I ask an obvious, impertinent question. The development plan sounds phenomenally exciting – but so did grand pronouncements from countless other fusion researchers over several decades. Why should we believe this is really it? 'You can find quotes from, like, the day before the Wright brothers flew, saying: "We're never gonna fly – we've been trying for so long already!"' Bob replies. He knows that it will take a full-scale operating model to convince the world at large – even as simulations using supercomputers have convinced him and his team that a historic breakthrough is finally in sight.

Within Bob's growing team, a striking number have come from SpaceX, the Elon Musk rocket business that recruited some of the world's most brilliant and ambitious mechanical engineers. Their migration reflects a shift in mindset that has swept the ranks of the US's highest performers in a wide range of fields, Bob says. 'If you're a really good engineer, or a really good scientist, or a really good lawyer, or anyone who wants to have an impact in the world – you go and work in climate, because that's the existential problem of anyone under forty,' he says. 'Going to Mars, that's one thing. But we have a planet here we gotta work on.'

Across the world's most powerful economy, some of its most brilliant innovators and wealthiest investors are scrambling for a leading role in the clean energy revolution. And while that revolution might stem in large part from the labs and boardrooms of a high-flying elite, it is already transforming lives in places that could scarcely be more different.

CHAPTER 22

With a climate crisis looming, consumers have the right to demand that products marketed as the ethical choice really stand up to scrutiny . . . The energy solutions of the future must not be based on the injustices of the past.

– Kumi Naidoo, secretary general of Amnesty International, 21 March 2019

KOLWEZI, DEMOCRATIC REPUBLIC OF CONGO

By all accounts the Kasulo cobalt rush started with a toilet. Between the unpaved lanes of this down-at-heel residential quarter in the southeastern Congolese city of Kolwezi, a man was digging a pit latrine behind his home when his shovel hit a crumbly black deposit, standing out against the ochre hue of the surrounding earth.

His timing was superb. For the first decades of mining in Congo's southeastern region, cobalt had been a virtual afterthought, dismissively referred to as a by-product of copper extraction. Compared with copper, which forms the skeleton of every electrical network in the world, cobalt – while important in the production of jet engines and artificial hips – was far less lucrative. Then, from the 1990s, came the adoption of lithium-ion batteries in the booming consumer electronics market. Cobalt was an indispensable component, interlayered with lithium in the positive electrode to maintain the battery's stability. Without it your smartphone would be a miniature firebomb. But the tiny size of phone batteries looked like a limiting factor for growth in cobalt mining – until the industry started paying serious attention to the growth of the electric car sector.

A single Tesla Model S battery contained more cobalt than a thousand iPhones. As Wall Street analysts started predicting that a growing focus on climate action would bring an explosion in the market for electric cars, the demand for cobalt looked set to mushroom. And most of the world's known reserves were in Congo, specifically in a strip of territory – a mere sliver of a country the size of western Europe – surrounding the dusty town of Kolwezi.

The toilet digger, whose name differs in every version of the story you hear in Kolwezi, tried to hide his discovery from prying eyes. Instead of digging further with the latrine hole, he sank a new one inside his house, carving down and sideways until he made contact with the same black seam, loading the ore into sacks and lugging them into minivans in the middle of the night, for dispatch to the ramshackle depots of Musompo, a market run by Chinese traders on the fringe of town. But Kasulo was a crowded, intimate community where bizarre behaviour was swiftly noticed. The secret got out – and found its way to a 12-year-old orphan named Laeticia.

It was just three weeks since the minibus crash that had killed her mother and left Laeticia badly injured, with a left foot splayed ninety degrees from its normal orientation. She'd moved into the cramped home of an uncle willing to provide what help he could, but clearly unable to afford her school fees. Then her friend Gisele, four years older, came up with an idea that would set Laeticia's childhood on a new course. 'Come with me tomorrow,' she said. 'We're going to Kasulo.'

The next morning the girls boarded a minibus packed with other children, all heading to that previously unremarkable neighbourhood on the northern edge of town. Limping through its muddy streets, Laeticia passed men dragging bicycles weighed down by hessian sacks that bulged with chunks of ore. She was jostled by hawkers with plastic boxes on their heads filled with sandwiches and deep-fried treats, announcing their presence to the miners with hisses and yells. She saw the subtler invitations of the prostitutes who had descended upon Kasulo alongside young men from across the region, sucked in by the mineshafts that were now proliferating across the neighbourhood like an infestation of woodworm, as residents rushed to turn their homes into mining sites.

Most inhabitants continued to live in their small brick houses, even as the ground beneath them became a honeycomb of pits and tunnels, and furious arguments broke out among them. Some had refused to jump on the cobalt bandwagon and were aghast at the destruction of the neighbourhood, as houses were degraded by cracks and fissures, and in some cases collapsed altogether. Scurrying amid the tumult were hundreds of small children, both Kasulo kids and others like Laeticia drawn from other parts of town by the chance to make money. Some of the boys found their way down the mines, but there was no way any of the teams of *creuseurs*, or diggers, would let a female join them underground, adult or otherwise, so Laeticia set to work as a scavenger. The scene was so chaotic, the earth so littered with fragments hacked from underground, that it was easy to sneak into mining areas and pluck bits of ore. Gisele taught Laeticia how to pick through the rocks in search of the telltale black streaks, how to break pieces of earth and clean the dirt from the cobalt. Discovery by the owner of a plot usually meant a hasty flight, but the more sympathetic ones could be placated with a packet of cigarettes.

Those first days were thrilling for Laeticia. Prowling Kasulo's mining sites, she could earn up to 2,000 francs a day – less than $2, but enough to keep herself fed without testing her uncle's generosity. The novelty soon wore off as she came to recognise the work for what it was: tough, unsafe drudgery, sickeningly unsuitable for a slight, injured child who should have had years of schooling ahead of her. Sometimes she was robbed or defrauded by the merchants who wandered the streets buying small amounts of low-grade ore from child scavengers. But the fees put school out of the question and she had to find some way to pay for food. So she kept working there for most of what remained of her childhood, her hands coarsened by the rocks she picked through, seared in the dry season and drenched in the rains, as Kasulo continued to draw profiteers and adventurers and desperate poor from across southeastern Congo to the giant ant nest taking shape beneath its floor. But a crunch was looming for this neighbourhood bonanza, heralded by the arrival in Kasulo's dirty streets of a rotund man in an expensive suit – one of the most powerful courtiers of President Joseph Kabila.

* * *

When King Leopold II of Belgium took personal possession of the Congo in the late nineteenth century, he went to extravagant lengths to present it as a philanthropic exercise aimed at protecting Congolese villagers from the predations of Arab slavers. In unguarded moments, however, he could let slip the real driving motivation. 'I don't want us to miss an opportunity to get ourselves a piece,' he wrote in one private letter, 'of this magnificent African cake.' The continent was there to be consumed – its bountiful resources and, in the process, vast numbers of its people, too.

After the invention of tyres made from rubber, Leopold's Congo became the world's leading supplier of the commodity, produced through brutal quota systems that led to the mass murder of dissenting villagers. Soldiers were required to sever hands to prove they had killed in sufficient volume, and sometimes took those of the living to make up the numbers. Such brutality and suffering was intrinsic to the supply chain that created the first petrol-powered cars in the early twentieth century. And the presence of children like Laeticia, amid appalling conditions, at the heart of the supply of a commodity vital to electric cars, was seized upon by modern human rights groups with a vehemence matching that of the campaigners against Leopold's crimes a century before.

Such controversy would be dangerous for any industry, but especially so for makers of electric vehicles, where the ethical credentials have been a key attraction for early adopters. BMW said it would start powering its electric cars with cobalt from Australia and Morocco. Elon Musk tweeted that the next generation of Tesla batteries would use a different technology requiring no cobalt at all. Congo's cobalt reserves have given this country, one of the largest and poorest in the world, a chance to play a central role in one of the twenty-first century's boom industries. But the dire reputation of its mining sector risks sabotaging all of that.

On the shore of Lake Nzilo an hour's drive east of Kolwezi, in an air-conditioned seafood restaurant owned by his niece, shod in crocodile-skin loafers and toying with a flute of Laurent-Perrier champagne, Richard Muyej is explaining to me his plan for tackling the crisis and preserving Congo's place in the green technology landscape. As the governor of Lualaba – the most crucial province to Congo's mineral-powered economy – Muyej is one of the most powerful men in the nation. He started out as a high school history teacher in the 1980s under the increasingly dysfunctional and kleptocratic thirty-year rule of Mobutu Sese Seko, who built palaces of stunning opulence in the jungle as the country lapsed into economic catastrophe. Mobutu was finally swept from power in 1997 by the rebel leader Laurent Kabila, who was assassinated four years later and succeeded by his 29-year-old son Joseph. Muyej, who had started moving into politics amid the chaos of Mobutu's last years, became a founding member of Joseph Kabila's new political party and his rise under the young leader was swift. In 2012 he was appointed interior minister, charged with maintaining the state's grip on a country the size of France, Germany, Poland, Iraq, Bangladesh, Japan and South Korea combined – with less than a tenth as many miles of paved road as Wales.

Within Congo's vast patchwork, the southeast has had a special place throughout its independent history. It has far more mineral wealth than any other part of the country – something that has fuelled a succession of attempts to break away and form an independent state. Keeping control of the region, and especially of its Lualaba province, where the bulk of the minerals are concentrated, has been one of the most vital missions for every Congolese government since the Belgian handover of 1960. And so, a couple of years after that innocuous latrine pit kicked off the Kasulo cobalt rush,

Interior Minister Muyej arrived in the neighbourhood to start imposing order.

Many residents remember angry *creuseurs* hurling stones in Muyej's direction; he insists that didn't happen but concedes they were furious. 'They shouted, "We're not on public land; we're on private plots! Let us do what we want on our own plots!" But what's under the ground doesn't belong to individuals. It belongs to the state.' Soon after that fraught visit, Muyej took over as governor of Lualaba and set about tackling the anarchy in Kasulo, which was undermining the image of the entire country among vital customers for its mineral exports. Over the course of a few months in 2017, Muyej oversaw the compulsory displacement of a whole section of Kasulo, which was fenced off and taken over by the Chinese cobalt giant Huayou. Homeowners were each given several thousand dollars in compensation – a large sum in Congo but far from enough to satisfy people who had counted on a rich stream of cobalt earnings for years to come.

Charities such as US-based PACT have helped to build a new future for many of the children who used to work in the area, including Laeticia. Now 17, she told me her story in the small tailoring workshop, perched by a busy highway, where she's now working as an apprentice. Behind a finely aged black Singer sewing machine, to the grumbling backdrop of passing trucks bearing minerals, she's earning enough to put her younger brother through school, and plans to open her own business in due course.

Still, not even Muyej claims that children have been entirely eliminated from the many 'artisanal' mines that still dot his province. A few days after our meeting, a lawsuit will be filed by US activists against some of the biggest global users of lithium-ion batteries, including Tesla. The suit, brought on behalf of the families of killed and injured child workers, accuses Tesla and the other defendants of 'abetting the cruel and brutal use of young children' in Congo's cobalt industry.

Most would agree it's unacceptable that the technology driving modern civilisation forward on a supposedly clean, sustainable path should involve dangerous work by children like Laeticia. But child labour is just the most vivid symptom of a far larger scandal: a corruptly mismanaged economy that has left this country's people among the world's poorest, as a tiny elite and their foreign partners siphon off the mining spoils. Far fewer children like Laeticia would have ended up at mine sites had the state not abjectly failed to provide free schooling as required by law. While some of Congo's children toil illegally underground, many thousands of others die each year from diarrhoea and other treatable ailments – victims of obscene corruption and negligence in a country that, despite its vast mineral wealth, has the lowest per capita health spending in the world. If international pressure succeeds only in shrinking the informal mining sector, replacing it with mines owned by well-connected foreign corporations, it will do little to ensure that Congo's people receive their fair share of the surging cobalt revenue.

While he avoids specifics, Muyej is surprisingly frank about the corruption that has cursed his nation's recent history. It's no use blaming foreigners, he says, for the value they have extracted from Congo through cushy mining deals. The 'really scandalous' thing, he says, has been the corruption involved, 'the absence of moral probity' on the part of officials who conspired to defraud the people of Congo and share the loot in secret. Even as a conspicuous beneficiary of the unequal division of spoils under the Kabila regime, which now sees him swigging premium champagne while a private jet waits to carry him to Kinshasa, Muyej doesn't bother feigning ignorance of its disastrous failings. And even as he fights to ensure that Congo reaps the rewards of the green tech explosion, he is darkly conscious of the potential for Lualaba's cobalt boom to deepen the mineral addiction that has had such pathological consequences for the country's governance. 'The mines,' he says, 'have drugged us.'

* * *

A hundred paces from the Kasulo market, where women sit at stalls selling used shirts and trousers for a dollar apiece, between a tall tree with thick green foliage and a tiny two-room house with tin roof and mud floor, a gang of small children has gathered around a hole descending deep into the ground, peering over its rim, where a coarse rope is lashed to a plank that spans its 5-foot diameter. Their voices, and the music swimming from someone's radio in a nearby home, recede to a dim murmur and then silence as I descend, holding my weight with stiffening fingers wrapped around the rope and with my bare toes gripping the crumbling terracotta interior of the shaft.

Beneath me Mangovo, a Barbie-doll-pink head torch strapped to his wet forehead, points out footholds invisible to the untrained eye, sometimes just grabbing one of my feet in a large coarse hand and shoving it into place. Occasionally, a piece of earth breaks and falls under my foot, adding a new burst to the floating dust whose smell mingles with that of our sweat in this airless space, whose temperature seems to rise with every metre we descend. At last, 40 feet underground, we squat at the base of the shaft on a floor carpeted with small fragments of stone and gaze up at the bright halo of sky far overhead, as distant as the light at the wrong end of a telescope.

We slither on our bellies through a small opening in the wall into a series of chambers just tall enough to sit up in. On one side, the *creuseurs* have stacked sandbags up to the ceiling to guard against sudden collapses. I wonder how much protection that offers as Mangovo grabs a chisel and starts stabbing at the black seam that runs like a river of spilled ink along the pale grey ceiling. He extends a lean arm towards me, its sinews honed by years of heaving himself up and down mine shafts and hacking at underground rock faces. Nestled in his palm is a fragment of a dull black lump that might pass for charcoal to the uninitiated. 'Cobalt,' Mangovo says. 'Nine per cent purity.'

Now in his forties, Mangovo has been a *creuseur* since the outset in the mid-1990s, when the state mining company virtually collapsed and unemployed men flooded into its disused pits. A Kasulo native, he was digging in the area around the first discovery until Muyej cleared it. Like many others, instead of abandoning Kasulo, he simply started digging in a different part of the neighbourhood outside the fenced-off section. It's the best work available to him, and the only work he knows. But it's *une vie médiocre*, he says as we sit in the subterranean darkness, a gruelling means of pursuing a decent future for his nine children.

The pit and the one next to it belong to Monique Swila, the *patron* who has been funding their construction and lives right beside them to keep an eye on the work. Every morning around seven, her little team of miners gathers in the covered porch outside her tiny shack, sitting at low tables and benches and smoking as they wait for her to serve them the staple southeastern Congolese meal: fried fish and vegetables with foufou, a mashed potato-like preparation from cassava flour. Even at this hour it's a rowdy atmosphere, the miners washing down their breakfast with local moonshine. Small children wander in and out, eavesdropping on the men's conversation before returning to their games on the edge of the unprotected mineshafts.

Monique, a strong character with broad, calm features and hair braided in cornrows, has no problem keeping the miners in check, having earned their respect by paying their expenses while they were sinking the shafts. They trust her to carry the ore they hack from the earth to the market at Musompo, and to give them their fair share of the earnings. She still dresses sharply, in bright prints and gold jewellery, as though for the streets of her native Lubumbashi, a heaving city of nearly two million, where she ran a clothing boutique with her husband until their relationship broke down. She turned to small-time trading to rebuild her capital, shuttling between Lubumbashi and the countryside. Then, two years ago, she leased this bit of land from Papa Jean, its 90-year-old owner, and set up as one of Congolese mining's innumerable micro-tycoons.

In the smoky darkness of her porch, Frank, the most charismatic member of her team, tells me they're days away from hitting a bumper seam that will unleash a flood of income. From the amused scepticism with which Monique looks on, it's clear that he's been giving this forecast for months. But she's as hungry for the big windfall as the rest of them. Cobalt offers her the quickest path to re-establishing herself in the clothing business. As soon as she's earned enough to set up a new boutique, she's heading back to Lubumbashi, to live in a house on a proper street with a private bathroom.

Business has been harder in recent months, the miners tell me, after the hitherto booming price of cobalt abruptly went into reverse. In part that's a reflection of their own productivity – Congo's cobalt production surged more quickly than that of the electric cars it was intended for, and a glut of the mineral built up in China. But Papa Jean's plot remains a hive of activity, with six more pits beside Monique's two. The old proprietor, who takes a third of the earnings from the mines, quietly potters around his devastated landholding in a threadbare grey suit jacket that swims around his shrunken shoulders, manifestly enjoying the excitement in his tenth decade.

Over the deepest pit, which extends a hundred feet underground, an orange tarpaulin has been erected against the rain, and a long tube extends down from an electric air pump rigged to a small generator, to stop the miners suffocating. On the surface, teams of men tug at a rope consisting of many shorter ones knotted together to haul up rock-filled sacks that once held Zambian maize, each weighing as much as a man.

Outside the pit, Guélord Selenge is preparing to climb down. Having helped to manhandle the sacks of ore from the previous shift, his skin is grey with dust and it's impossible to determine the original colour of his work clothes. Aged 31, he's been suffering the impact of the rollercoaster price fluctuations, but sees no better career opportunities. 'The problem is you foreigners,' he says. 'You come and buy our minerals cheap, close off the sites so we can't enter, and even if we do manage to get some ore we have to sell it to you for a low price.' He pauses. Someone is calling him – it's almost time to go underground. '*Le problème*,' he repeats, '*c'est vous*.'

One day Monique takes me with her on a trip to sell cobalt at the market. She's looking her sharpest, with a small gold dollar sign suspended on a pendant necklace. The miners heave sacks of ore into the back of a beat-up yellow minivan, and we squeeze into the broken front seat. We bounce through the rutted streets of Kasulo, gluey from overnight rainfall, before turning on to the northbound highway that will take us to Musompo.

The market district announces itself with a string of long, flat sheds on either side of the road, behind walls emblazoned with the names of the proprietors, some using their Chinese names – Boss Wu, Mr Liu – others adopted European ones like Yannick or Sarah. From these rickety installations, cobalt ore will be transported in lorries to plants in Lubumbashi run by larger Chinese companies such as Huayou. The cobalt is refined and then sent overland to Mombasa and Durban, whence it is shipped to southern China and further processed for use in cells made by the giant Chinese, Japanese and South Korean battery makers, before assuming its final incarnation deep inside the world's growing fleet of electric cars.

We pass a couple of dozen depots before turning left onto a street lined with women sitting with vegetables laid out on strips of tarpaulin. A sharp right takes us to a depot smaller than the rest, perched on a quiet side street, where Ke Xiaote is in charge. With round, thin-rimmed spectacles and a shock of gelled hair, Xiaote's demeanour contrasts with those of the other Chinese merchants I encountered on a previous trip to Musompo. Where they tended to be withdrawn, older men silently chain-smoking as they watched their Congolese employees handle negotiations, Xiaote has the permanent half-smile of a man on an adventure as he banters with the Kasulo cobalt sellers in his broken Swahili. Aged 25, he's been in

Congo since 2016, when he followed his brother here after finishing a computer science degree in Guangzhou. We speak amid the crashing of cobalt ore being crushed with mallets, one of his employees serving as the conduit between his awkward Swahili and my awkward French. He goes home for just one month a year and has no plans to leave this business, he says. You must like it here, I say. He wrinkles his nose and switches to English as he gestures at a black mound of dusty ore on the concrete floor. 'I like this,' he murmurs.

When his men have crushed a batch of ore into a coarse powder, Xiaote takes a sample for testing on his Metorex, a handheld device that measures mineral purity. On the wall over his shoulder is a list showing the price paid for the ore depending on the cobalt content. The Metorex has become hated among the *creuseurs*, who are convinced that the merchants have conspired to rig their measuring devices in order to rip off local sellers. Muyej himself raised a laugh at an investment conference I attended, by gesturing at his ample gut and claiming his weight would look far healthier on one of the Chinese scales at Musompo.

Compared with the other Chinese merchants, who are routinely condemned by *creuseurs* as thieves, Xiaote has a reputation for relative decency – perhaps helped by his practice of giving gifts to regular sellers. After a spirited negotiation with Xiaote, in which at one point she jumps into the throne-like chair from which he likes to survey the room, Monique leaves smiling with 175,000 francs and a new red dress in a plastic packet. Most of the cash will go to the miners and Papa Jean, but her share still amounts to about $30 – roughly the average Congolese monthly wage, from just a couple of days of production, despite the recent price decline. Monique's plan of rebuilding a respectable life in Lubumbashi remains on track. Her takings from the mine will surge when Frank finally finds his promised seam, and it's just a matter of time, she's sure, until accelerating electric car demand clears the market glut and sends the cobalt price shooting back up.

But her exit from Kasulo may come sooner than she realises. At my meeting with Governor Muyej, he told me the scale of his plans for the district. It's not feasible to eliminate informal artisanal mining overnight, he said – but if Congo is to preserve its place in the clean tech supply chain, there's no place for the likes of Monique, running hazardous mining operations that create tunnels under houses amid scrappy bits of land where children play. He wants to clear the whole of Kasulo to be handed over to an industrial mining group. It will cost about $800 million. As soon as the cobalt price recovers, he hopes a cash-flush minerals company will be ready to take the plunge.

Until then, Monique is set to stay in business. As the sky starts to darken over the Musompo depots, she uses some of her earnings to buy us some sliced mango from a hawker with a baby on her back, and we squeeze into another minibus with broken seats, packed with traders and miners heading back to Kasulo. The radio is blaring out a song by Papa Wemba, the late hero of Congolese rumba, with his unmistakable blend of melancholy and defiant energy, and Monique sings along as she leafs through Xiaote's yellow 20,000 franc notes – her slice of the spoils of a greener future.

Afterword

I'm a reporter, not an activist or a scientific expert, and in this book I've tried to let the stories speak for themselves. There's no shortage of books offering blueprints for tackling the climate crisis – some of them excellent – and this isn't intended to be one of them. In closing, however, I want to mention a few things that struck me particularly powerfully during my research, and which I'll be following closely in the years ahead.

Humanity won't be driven to extinction by climate change, just as our prehistoric forebears survived environmental shifts from the start of our evolutionary history. Large numbers of us will respond much as early humans did to the waxing and waning of ice ages: by moving. If that sounds reassuring, it shouldn't. When the last ice age ended more than 10,000 years ago, the global population was at most a few million, all of them hunter-gatherers. The world was a thousand times less densely populated than today, in other words, with no land ownership, no national borders, no cities or villages or even farms. Our ancestors migrated gradually over centuries in response to a natural, cyclical temperature rise that was just one-tenth the speed of the pollution-driven warming we have now unleashed.

As it crams people into overcrowded cities like Dhaka, or tears apart communities like Barangay 48, the climate migration of the twenty-first century is shaping up to be something far more traumatic. Most of the movement will take place within developing countries, which will be disproportionately affected by climate change due to their largely tropical geography and largely agrarian economies. But many of those displaced will seek to rebuild their lives abroad. For a sense of the butterfly effect this could set off in the rich world, we can look to the civil war in Syria, which followed its region's worst extended drought on record. That conflict drove a large flow of refugees towards Europe, which was exploited to powerful effect by right-wing nationalists from England to France to Hungary. The potential for still uglier strains of politics to flourish in response to a vast and growing wave of climate refugees should be obvious.

Heading off domestic extremism is one reason for rich countries to invest heavily in helping developing nations adapt to climate turmoil. Another is that, as Robert Tehena pointed out in the Solomon Islands, there's no ethical justification for not doing so, given the rich world's overwhelming contribution to this crisis. For each of its inhabitants, the United States emits 15.5 tonnes of carbon a year, compared with 0.1 tonnes for Ethiopia. Even China, after its massively polluting industrial growth of recent decades, has still put much less carbon in the atmosphere than either the US or the EU, with their far smaller populations. Yet climate-related assistance is still falling far short of requirements, as is international aid in general. All but a handful of rich countries are failing even to meet the target, agreed at the UN over fifty years ago, of spending 0.7 per cent of gross national income on overseas assistance. The Covid-19 pandemic did nothing to change this – indeed the UK, one of the few countries that had been hitting the target, cited the global crisis as justification to abandon it.

In other ways, however, the response to the pandemic – which broke out midway through my work on this book – gives grounds for hope on climate action. It has exposed the fragility of the stable, prosperous existence that many of us had taken for granted. It's been a reminder to pay attention to scientists. Most striking, for me, was the public support for unprecedented restrictions to slow the virus's spread. After European nations imposed their lockdowns, opinion polls showed overwhelming approval, including among young people who faced relatively slight personal danger from the coronavirus. Whatever you might think of the lockdowns themselves, this was a devastating blow to one of the most insidious arguments against serious climate action: that it's not realistic, because the citizens of modern society are just too selfish to make the sacrifices involved. In fact, we now know, people are willing to undergo extraordinary levels of disruption to their lives if they're convinced it's ethically required.

And compared with the lockdowns, the disruption to most people's lives from a rapid energy transition will be tiny. In a world without alternatives to fossil fuels, we would be facing a nightmarish choice between economic shutdown and ecological catastrophe. We've been spared that fate by the stunning recent advances in green technologies. Some, like solar power and electric cars, are already cheap enough to compete in the open market. Others, like Sunfire's synfuels, still need further investment to roll out at scale and bring down costs. But it's clear that, as Edda Aradóttir put it to me in Iceland, we now have all the basic technologies we need to tackle the climate crisis. The next challenge is to make sure we develop and use them – and to kill the financial incentive to just keep burning fossil fuels instead. That means sweeping reform of some of the fundamental pillars of the global economic framework. And there's a growing realisation, at the highest levels of politics and business, that the overhaul is overdue.

'Our political and economic systems are geared towards a world where the damage caused by carbon emissions was not understood,' Hank Paulson, treasury secretary under George W. Bush, told me during the US leg of my research. For Paulson, a former Goldman Sachs CEO, the answer is to unleash the power of the market, by putting a hefty price on emissions. That concept has been resisted by some in the environmental movement, who warn that it was market-based systems that got us into this mess. Personally, I think a massively expanded carbon pricing framework could be the single most powerful tool available to policymakers in the global response to climate change, punishing the worst polluters while making green technologies more cost-competitive. And if the fastest-moving countries start to apply the carbon price to imports – as Frans Timmermans' team is aiming to do in the EU – it will give other nations a powerful incentive to fall in line.

As I write this, global corporations are falling over each other to publish plans to bring their net emissions down to zero, in anticipation of government measures imposing heavy costs on companies that fail to do so. It remains to be seen whether those measures will materialise at the level required. Governments, especially in the Western democracies, will take radical action only if they're convinced that their population has the stomach for it. Voters who push leaders to follow through on lofty rhetoric will not only be doing the right thing for those most vulnerable to climate change, but also setting up their own nations to prosper in the post-fossil fuel age.

It's impossible to predict how this race will play out, as the effects of climate change continue to ripple through the planet and the global economy. What is certain is that this will remain the biggest and most important story that I, or any journalist of my generation, will get the chance to cover. And it's just beginning.

Acknowledgements

During my research for this project I spoke to many hundreds of people, all of whom contributed in a huge range of ways to my understanding of the subjects covered. To all those mentioned by name in the preceding pages – and to the many more who are not – my sincerest thanks. To state the blindingly obvious, this book could not exist without you.

It would have been similarly impossible to write it without the assistance of the guides, interpreters and research partners who accompanied me on many legs of the journey. Valentin Struchkov was a superb companion from Yakutsk to Tiksi, professional and enthusiastic throughout our Siberian travels, even as we shivered on our long boat ride to and from the tusk hunters' camp. My thanks also to Erel Struchkov for sharing his knowledge on Batagaika and the tusk hunting there. 'Superguide' Lakshmi Kumar Rai was a stalwart presence throughout our trek to Tsho Rolpa and the winding journey to Tatopani. In Greenland, Paarnannguaq Andersen, Aleqatsiaq Peary and Rasmine Jeremiassen Normann did a great job of interpreting in my interviews in Ilulissat and Qaanaaq. James Wada steered me through the waterways of Makoko and introduced me to a compelling range of residents. James Taluasi was a sterling companion in Fanalei and Malaita, and spoke eloquently about the situation for my video on the subject. Shohail Bin Saifullah and Syed Siyam provided tremendous assistance during my fieldwork in Dhaka, as did Kazi Abrar Tarik on our frantic trip across the south-west. Pratibha Shinde was a highly capable and diligent interpreter throughout my visits to Tulsiram and other farmers in Jalna. Chuluunbat and Batkhishig were terrific partners on the winter journey across Mongolia, and fine instructors in the technique of digging jeeps out of snowdrifts. I could have had no better companion in the Amazon than Marco Lima, a true expert on the rainforest and a tireless collaborator in my work there. In Congo, Ephraim Mununga Tshisola was a top-notch Swahili interpreter and helped me track down mining operations in Kasulo.

I was extremely fortunate at several points on my travels to meet people who saw value in this project and were hugely generous in their support for it. My research in Chile owed a vast amount to Marcelo 'Reta' Retamal, a giant of the country's wine sector, who showed remarkable kindness in introducing me to winemakers along the length of the country. Valerie Browning and her charity APDA are doing phenomenal work to address the serious problems facing the people of Afar, and she and her colleagues were of tremendous help in my research there. In Ethiopia's Somali region, President Mustafa Omer and Hassan Farah were greatly supportive, and Abdulfatah Osman Abdi was a valued companion on the journey to the region's south. In the Solomon Islands I was lucky to meet Maclean Biliki of the Lauru Land Conference of Tribal Community, who accompanied me to Nuatambu and gave invaluable support on that trip. Mark Hutchison and Julie Hollis were wonderfully hospitable in Nuuk, and sources of guidance both practical and geological. In Venice, I benefited from the knowledge and insights of Jane Da Mosto, the founder of We Are Here Venice, as well as her colleague Eleonora Sovrani's support in my research. In Nepal, Dhananjay Regmi generously shared his expertise in helping me plan my trip to the mountains.

I benefited throughout this project from the insights and advice of fellow journalists at the FT and elsewhere, particularly the following, to whom I hope to repay the good turn before long: Ahmed Al-Omran, Tom Burgis, Will Clowes, James Crabtree,

William Davison, Tom Gardner, Kathrin Hille, Lucy Hornby, Michael Kavanagh, Amy Kazmin, Daniel Knowles, Fabiano Maisonnave, David Pilling, Oliver Ralph, Sophy Roberts, Jancis Robinson, Henry Sanderson, Susannah Savage, Andres Schipani, Jamie Smyth, Dominic Standish, Tom Wilson, and Yuan Yang.

My agent Sophie Lambert believed in this project from the very start, and was full of wise advice as I developed the concept and the material. Jo Thompson was a terrific editor, with vast enthusiasm and razor-sharp instincts. Alex Gingell expertly shepherded the manuscript through to publication, and it was a pleasure to work with Helen Upton on publicity and Lindsay Terrell on marketing. Emma Pidsley did a wonderful job on the cover design, as did Richard Collins and Anne O'Brien on copy editing and proofreading respectively. Thanks also to Anurag Banerjee for the portrait photograph in Mumbai.

I'm grateful to all the friends who provided moral support and sounding boards for the ideas in this book – especially Shilpa Rathnam, Tim Chase, Tom Graham and Daniel Fogg. And of course to my family – my sister Charlotte, whose literary insights were invaluable throughout the process, and my parents, who have never wavered in their encouragement for this or any other of my far-flung escapades.

Thank you all.

List of Illustrations

All photographs were taken by the author and are copyright of Simon Mundy. For more photos and videos of his journey, visit simonmundy.com.

Notes and References

CHAPTER 1

7 *Its growth was turbo-charged from the 1990s onward*: V. Vadakkedath et al., 'Multisensory satellite observations of the expansion of the Batagaika crater and succession of vegetation in its interior from 1991 to 2018', *Environmental Earth Sciences* 79, 150 (2020).

7 *a Russian permafrost zone that is the size of* . . .: S. Gruber, 'Derivation and analysis of a high-resolution estimate of global permafrost zonation', *The Cryosphere*, 6, 221–33 (2012); World Bank data retrieved from https://data.worldbank.org/indicator/AG.LND.TOTL.K2

8 *Beneath Siberia's frozen soil lies billions of tonnes . . . microbes are feasting* . . .: E. Schuur et al., 'Climate change and the permafrost carbon feedback', *Nature* 520, 171–9 (2015).

8 *carbon emissions from the Arctic permafrost are*: T. Schuut, 'Permafrost and the Global Carbon Cycle', National Oceanic and Atmospheric Administration: *2019 Arctic Report Card* (2019).

8 *ensure the permafrost kept thawing, releasing still more gas*: A. H. MacDougall et al., 'Significant contribution to climate warming from the permafrost carbon feedback', *Nature Geoscience* 5(10), 719–21 (2012).

8 *-64°C in Yakutsk*: A. Shver and S. A. Izyumenko. Климат Якутска. Leningrad: Ленинград Гидрометеоиздат (1982).

8 *-68° in Oymyakon*: A. Clarke, 'The thermal limits to life on Earth', *International Journal of Astrobiology*, 13(2), 141–54 (2014).

8 *an expanding blanket of dark seawater*: A. Riihelä et al., 'Observed changes in the albedo of the Arctic sea-ice zone for the period 1982–2009', *Nature Climate Change* 3, 895–8 (2013).

8 *The global average temperature rose by more than 1°C*: Data from NASA's Goddard Institute for Space Studies (GISS). Retrieved from https://data.giss.nasa.gov/gistemp/graphs/graph_data/Global_Mean_Estimates_based_on_Land_and_Ocean_Data/graph.txt

8 *half a degree per decade*: A. Gorokhov and A. Fedorov, 'Current Trends in Climate Change in Yakutia', *Geography and Natural Resources* 39(2), 153–61 (2018).

9 *more than the average Russian monthly salary*: CEIC data retrieved from https://www.ceicdata.com/en/indicator/russia/monthly-earnings

9 *Engraved mammoth tusks can fetch more than $1 million*: J. Starkey, 'Mammoth tusks on sale for up to £2.3m', *The Times*, 17 July 2015.

13 *being published in* Science: S. A. Zimov, E. A. Schuur and F. S. Chapin III, 'Permafrost and the global carbon budget', *Science* 312(5780), 1612–13 (2006).

15 *Church claims . . .*: R. Anderson, 'Welcome to Pleistocene Park', *The Atlantic*, April 2017.

15 *competition from South Korea's Sooam Biotech*: 'Three party collaboration between the Korea, China and Russia for the restoration of the mammoth', Website of Sooam Biotech: http://en.sooam.com/html/?code=B01 (Retrieved May 2021).

CHAPTER 2

19 *335 billion tonnes a year*: M. Zemp et al., 'Global glacier mass changes and their contributions to sea-level rise from 1961 to 2016', *Nature* 568, 382–6 (2019).

20 *more than 5,000 lakes*: G. Veh, O. Korup and A. Walz, 'Hazard from Himalayan glacier lake outburst floods', *Proceedings of the National Academy of Sciences* 117(2), 907–12 (2020).

20 *NASA data shows*: D. Shugar et al., 'Rapid worldwide growth of glacial lakes since 1990', *Nature Climate Change* 10, 939–45 (2020); NASA press release: 'Global Survey Using NASA Data Shows Dramatic Growth of Glacial Lakes', 31 August 2020.

20 *Tsho Rolpa's 85 billion litres*: M. V. Peppa et al., 'Glacial lake evolution based on remote sensing time series: a case study of Tsho Rolpa in Nepal', *ISPRS Ann. Photogramm. Remote Sens. Spatial Inf. Sci.*, V-3-2020, 633–9 (2020).

20 *one researcher described to me*: Thanks to Milan Shrestha, a

Himalayan glacial lake specialist at Arizona State University, for this evocative analogy.

20 *a British–Nepalese team of researchers*: ICIMOD press release: 'Reassessing Tsho Rolpa glacial lake', published at https://www.icimod.org/reassessing-tsho-rolpa-glacial-lake/ (24 July 2019).

21 *minutes, at most*: K. C. Deepak, 'Early Warning System: a case study of Tsho Rolpa Glacial Lake', presented at UNDP Workshop on Disaster Risk Management in Hydropower (15 November 2017).

21 *$3 million project*: S. R. Bajracharya, 'Glacial lake outburst floods risk reduction activities in Nepal', paper presented at Asia Pacific Symposium on New Technologies for Prediction and Mitigation of Sediment Disasters, Tokyo, Japan (November 2009).

24 *99 per cent of Nepalese electricity generation*: 'Energy Sector' page on Nepal government portal: https://nepal.gov.np:8443/NationalPortal/view-page?id=92 (Retrieved May 2021).

26 *twenty times larger*: K. L. Cook et al., 'Glacial lake outburst floods as drivers of fluvial erosion in the Himalaya.' *Science* 362(6410), 53–7 (2018); W. Wang et al., 'Integrated hazard assessment of Cirenmaco glacial lake in Zhangzangbo valley, Central Himalayas', *Geomorphology* 306, 292–305 (2018).

26 *another four times*: Peppa et al. (2020).

CHAPTER 3

31 *around the start of the thirteenth century*: T. M. Friesen and C. D. Arnold, 'The Timing of the Thule Migration: New Dates from the Western Canadian Arctic', *American Antiquity* 73(3), 527–38 (2008).

31 *contact between the two groups*: H. C. Gulløv, 'The Nature of Contact between Native Greenlanders and Norse', *Journal of the North Atlantic* 1(1), 16–24 (2008).

31 *Norse settlements died out*: L. K. Barlow et al., 'Interdisciplinary investigations of the end of the Norse Western Settlement in Greenland', *The Holocene* 7(4), 489–99 (1997).

31 *covers four-fifths*: A. K. Rennermalm et al., 'Understanding Greenland ice sheet hydrology using an integrated multi-scale approach', *Environmental Research Letters* 8, 015017 (2013).

31 *vertical thickness of up to 2 miles*: J. L. Bamber, R. L. Layberry and S. P. Gogineni, 'A new ice thickness and bed data set for the

Greenland ice sheet: 1. Measurement, data reduction, and errors', *Journal of Geophysical Research* 106, 33773–80 (2001).

32 *drawn from a meteorite*: M. J. O. Svensson et al., 'Methods for determination of the source of iron in precontact Inuit and Dorset culture artifacts from the Canadian Arctic', *Journal of Archaeological Science: 17* 36, 102814 (2021).

32 *he charged 20,000 people 25 cents each*: D. Smith, 'An Eskimo Boy And Injustice In Old New York; A Campaigning Writer Indicts An Explorer and a Museum', *New York Times*, 15 March 2000.

32 *forced to abandon several settlements*: M. Takahashi, 'Greenland's Quest for Autonomy and the Political Dynamics Surrounding the Thule Air Base', in Takahashi, M. (ed.), *The Influence of Sub-state Actors on National Security*, 25–50, Springer Polar Sciences (2019); S. F. Gearheard et al. (eds), *The Meaning of Ice*, International Polar Institute (2013).

33 *ten days each decade*: J. Stroeve and D. Notz, 'Changing state of Arctic sea ice across all seasons', *Environmental Research Letters* 13, 103001 (2019).

33 *self-perpetuating feedback cycle*: H. Kashiwase et al., 'Evidence for ice-ocean albedo feedback in the Arctic Ocean shifting to a seasonal ice zone', *Scientific Reports* 7, 8170 (2017).

33 *triple the global average rate*: Arctic Monitoring and Assessment Programme (AMAP), *Arctic Climate Change Update 2021: Key Trends and Impacts* (May 2021).

33 *40 per cent thinner*: Stroeve and Notz (2019).

33 *the government imposed kill quotas*: J. Olsen, 'Global warming forces polar bear hunt quota in Greenland', Associated Press (23 February 2006).

33 *he was hunting near the ice edge*: This story was first reported in Gretel Erlich's wonderful account of her experiences with the Inughuit people: 'Rotten Ice: Traveling by dogsled in the melting Arctic', *Harper's*, April 2015.

34 *Brigitte Bardot led a campaign*: L. Rasmussen, 'The Wrong Seal Hunt', *New York Times*, 21 March 1979.

35 *a gunfight broke out*: '31-årig løsladt', *Sermitsiaq*, 17 July 2009.

35 *scores of valuable mineral deposits*: See the Greenland Mineral Resources Portal, available at http://www.greenmin.gl/

36 *Tiksi, an Arctic port built . . .*: D. Maximova, 'Sustainable Development of the Russian Arctic Zone: Challenges &

Opportunities', in L. Heininen and H. Exner-Pirot (eds), *Arctic Yearbook 2018*, 373–87 (2018).

36 *The population plunged*: 'Географическая и историческая справка', retrieved from the Tiksi government website: https://tiksi.sakha.gov.ru/ob-omsu-rsja/geograficheskaja-i-istoricheskaja-spravka/

36 *Putin had given his officials seven years*: 'Владимир Путин: Северный морской путь должен стать ключом к развитию Дальнего Востока', press release from Russia's Ministry for Development of the Russian Far East and Arctic (1 March 2018).

36 *the first container ship to complete the route*: 'Maersk 3,600-TEUer traverses Northern Sea Route, but icebreaker vital', Shanghai International Shipping Institute (17 September 2018); Louppova, J., 'Venta Maersk: first containership passed Northern Sea Route', Port.today (25 September 2018).

37 *proclaiming itself a 'near-Arctic' power*: S. Reinke de Buitrago, 'China's Aspirations as a 'Near Arctic State': Growing Stakeholder or Growing Risk?', in Weber J. (ed), *Handbook on Geopolitics and Security in the Arctic*, Springer (2020).

37 *Chinese state companies nearly sealed a deal*: D. Hinshaw and J. Page, 'How the Pentagon Countered China's Designs on Greenland', *Wall Street Journal*, 10 February 2019.

37 *'Essentially it's a large real estate deal'*: 'Trump likens buying Greenland to "a large real estate deal"', Associated Press (18 August 2019).

37 *cancelled a state visit to Copenhagen*: J. Mason and N. Skydsgaard, 'Trump calls Danish PM's rebuff of Greenland idea "nasty" as trip cancellation stuns Danes', Reuters (21 August 2019).

37 *the capital city of 18,000*: Data retrieved from StatBank Greenland at https://bank.stat.gl/

38 *Under a 2009 agreement with Copenhagen . . .*: E. Wilson, *Energy and minerals in Greenland: governance, corporate responsibility and social resilience*, IIED (2015).

39 *it deposits tens of billions of tonnes of ice*: K. D. Mankoff et al., 'Greenland Ice Sheet solid ice discharge from 1986 through March 2020', *Earth Syst. Sci. Data* 12, 1367–83 (2020).

39 *the most likely source of the iceberg that sank the* Titanic: R. Harris, 'Glacier Blamed For Berg That Sank Titanic Unleashes More Ice', NPR.org (3 February 2014).

39 *began losing mass at an accelerating pace*: J. H. Bondzio et al., 'The

mechanisms behind Jakobshavn Isbræ's acceleration and mass loss: A 3-D thermomechanical model study', *Geophysical Research Letters* 44(12), 6252–60 (2017).

41 *increased by 7.3°C*: E. Hanna et al., 'Recent warming in Greenland in a long-term instrumental (1881–2012) climatic context: I. Evaluation of surface air temperature records', *Environmental Research Letters* 7 045404 (2012).

41 *meltwater from the sheet's surface was trickling down*: H. J. Zwally et al., 'Surface Melt-Induced Acceleration of Greenland Ice-Sheet Flow', *Science* 297(5579), 218–22 (2002).

42 *between 26 and 110 centimetres*: IPCC. 'Summary for Policymakers,' p. 20, point B.3.1, in: IPCC Special Report on the Ocean and Cryosphere in a Changing Climate (2019).

43 *Three days after arriving at Swiss Camp . . .*: A. Ringgaard, 'Verdenskendt glaciolog død i tragisk ulykke på Indlandsisen', Videnskab.dk (14 August 2020); J. Schwartz, 'Konrad Steffen, Who Sounded Alarm on Greenland Ice, Dies at 68', *New York Times*, 13 August 2020; J. Berardelli, 'Greenland ice sheet claims life of renowned climate scientist', CBS News (12 August 2020).

CHAPTER 4

49 *twenty million . . .*: Lagos State Government, 'About Lagos.' https://lagosstate.gov.ng/about-lagos/

49 *the ocean ate away at Victoria Island*: P. C. Nwilo et al., 'Long-term determination of shoreline changes along the coast of Lagos', *Journal of Geomatics* 14(1) (2020).

50 *lethal inundations*: 'Flood sweeps 11-year-old boy, rescuer to death, destroys property in Lagos', Nigeria News Network (12 October 2019).

50 *set to become increasingly severe*: B. J. Abiodun et al., 'Potential impacts of climate change on extreme precipitation over four African coastal cities', *Climatic Change* 143, 399–413 (2017).

50 *clogged, antiquated drains*: A. Adeloye and R. Rustum, 'Lagos (Nigeria) flooding and influence of urban planning', *Urban Design and Planning* 164, 175–87 (2011).

51 *disembarked by mistake*: M. H. Y. Kaniki, 'The Psychology of Early Lebanese Immigrants in West Africa', *Transafrican Journal of History* 5(2), 139–47 (1976).

52 *funnelled about $4 billion of public funds*: I. Jimu, 'Managing Proceeds of Asset Recovery: The Case of Nigeria, Peru, the

Philippines and Kazakhstan', International Centre for Asset Recovery, Working Paper Series No. 6 (2009).

52 *money-laundering conviction*: Summary of case 1A.215/2004 published by Swiss Federal Tribunal in Lausanne on 7 February 2005. Retrieved from UNODC Sherloc database.

52 *gross national income works out at about $2,000 per person*: World Bank data retrieved from https://data.worldbank.org/indicator/NY.GDP.PCAP.CD?locations=NG

52 *report in* Rolling Stone *magazine*: J. Goodell, 'The Climate Apartheid: How Global Warming Affects the Rich and Poor', *Rolling Stone*, 24 October 2017.

53 *made an effusive congratulatory speech*: 'Eko Atlantic 5,000,000 Square Metres Dedication Ceremony', published by Eko Atlantic at https://www.youtube.com/watch?v=ULccxM3JaqA

53 *media homed in*: J. Tanfani, 'He was a billionaire who donated to the Clinton Foundation. Last year, he was denied entry into the U.S.', *Los Angeles Times* (28 August 2016).

56 *predicted that it would indeed shift the coastal erosion eastwards*: Eko Atlantic Environmental Impact Assessment, October 2012. Retrieved from website of Nigerian Environmental Assessment Department: https://ead.gov.ng/wp-content/uploads/2017/08/EKO-ATLANTIC-PHASE-I-Final-EIA-Report-Main-Report-October-2012.pdf

56 *sediments spat out by the mighty Volta River*: A. Blivi et al., 'Sand barrier development in the bight of Benin, West Africa', *Ocean & Coastal Management* 45, 185–200 (2002).

58 *migrants from the powerful Kingdom of Dahomey*: O. J. Koko, 'A study of etymology, ethnology and lexical hybrid in Ogu language', *Journal of Languages, Linguistics and Literary Studies* 6 (2018).

58 *the state governor said*: 'Makoko: Fashola explains demolition of shanties', PM News (23 July 2012).

59 *claimed the life of Timothy Huntoyanwha*: O. Akoni and M. Olowoopejo, 'Slain Lagos chief: Police assures of justice', *Vanguard*, 22 July 2012.

60 *Adeyemi won the coveted Silver Lion*: J. Mairs, 'Kunlé Adeyemi docks Makoko Floating School at the Venice Biennale', Dezeen.com (31 May 2016).

60 *Adeyemi had spent most of his career*: 'Kunle Adeyemi', Website of LafargeHolcim Foundation, retrieved May 2021: https://www.lafargeholcim-foundation.org/experts/kunle-adeyemi

60 *which was winning enthusiastic international media coverage*: For example: J. Collins, 'Makoko Floating School, beacon of hope for the Lagos "waterworld"', *Guardian*, 2 June 2015.

60 *rejected by the head of the local school*: A. Gaestel, 'Things Fall Apart', *The Atavist*, No. 76 (February 2018).

60 *the celebrated new structure crumpled*: S. Ogunleye, 'Floating school in Lagos lagoon collapses under heavy rains', Reuters (8 June 2016).

CHAPTER 5

66 *on this and countless other reefs across the Maldives*: N. Ibrahim et al., 'Status of Coral Bleaching in the Maldives in 2016', Marine Research Centre, Maldives (2017).

66 *global average temperature hit yet another modern record*: 'NASA, NOAA Data Show 2016 Warmest Year on Record Globally', NASA press release (18 January 2017).

66 *sea temperatures surged well above 30°C*: Data retrieved from NOAA Coral Reef Watch database at https://coralreefwatch.noaa.gov/product/vs/gauges/maldives.php

67 *providing crucial data for Charles Darwin's groundbreaking first book*: C. Darwin, *The Structure and Distribution of Coral Reefs*, Smith Elder (1842).

67 *Seismic mapping of the Maldivian atolls*: C. Betzler et al., 'Lowstand wedges in carbonate platform slopes (Quaternary, Maldives, Indian Ocean)', *The Depositional Record* 2(2), 196–207 (2016).

67 *is fish shit*: C. T. Perry et al., 'Linking reef ecology to island building: Parrotfish identified as major producers of island-building sediment in the Maldives', *Geology* 43(6), 503–6 (2015).

67 *the oceans rose by 120 metres*: J. D. Stanford et al., 'Sea-level probability for the last deglaciation: A statistical analysis of far-field records', *Global and Planetary Change* 79(3–4), 193–203 (2011).

67 *were formed from pieces of coral skeleton*: P. S. Kench et al., 'New model of reef-island evolution: Maldives, Indian Ocean', *Geology* 33(2), 145–8 (2005).

67 *first arrived about 3,500 years ago*: E. Knoll, 'The Maldives as an Indian Ocean Crossroads', in *Oxford Research Encyclopedia of Asian History*, Oxford University Press (2018).

67 *a sailor named Boyce wrote*: L. Vilgon, *Maldive ODD History: The Maldives Archipelago and its people*, Vol. 4, p. 129. Accessed

through digital repository of the Maldives National University: http://saruna.mnu.edu.mv/jspui/handle/123456789/951

68 *worst wave of coral bleaching*: C. M. Eakin et al., 'The 2014–2017 global-scale coral bleaching event: insights and impacts', *Coral Reefs* 38, 539–45 (2019).

68 *ocean has absorbed two thirds of the heat*: L. Zanna et al., 'Global reconstruction of historical ocean heat storage and transport', *PNAS* 116(4), 1126–31 (2019).

68 *increased its average surface temperature by about 1°C*: C. Deser et al., 'Twentieth-century tropical sea surface temperature trends revisited', *Geophysical Research Letters* 37(10) (2010).

68 *warned that global warming was on course*: O. Hoegh-Guldberg et al., Impacts of 1.5°C Global Warming on Natural and Human Systems, in *Global Warming of 1.5°C. An IPCC Special Report* (2018).

69 *support a quarter of the world's marine species*: O. Hoegh-Guldberg et al., 'People and the changing nature of coral reefs', *Regional Studies in Marine Science* 30 (2019).

70 *80 per cent of the country*: A. Voiland, 'Preparing for Rising Seas in the Maldives', NASA Earth Observatory. Retrieved at https://earthobservatory.nasa.gov/images/148158/preparing-for-rising-seas-in-the-maldives

70 *fleeing to exile in Britain*: 'Mohamed Nasheed: Former Maldives president "given UK asylum"', BBC News (23 May 2016).

70 *'not prepared to die'*: N. Chestney, 'Maldives tells U.N. climate talks: "We are not prepared to die"', Reuters (13 December 2018).

71 *By interbreeding these unusually hardy survivors*: M. Van Oppen et al., 'Building coral reef resilience through assisted evolution', *PNAS* 112(8), 2307–13 (2015).

71 *subjecting corals to near-death experiences*: E. Gibbin et al., 'Short-Term Thermal Acclimation Modifies the Metabolic Condition of the Coral Holobiont', *Front. Mar. Sci.* 5:10 (2018).

71 *At California's Stanford University*: P. Cleves et al., 'CRISPR/Cas9-mediated genome editing in a reef-building coral', *PNAS* 115(20) 5235–40 (2018).

72 *young German tourist sent here in 1978*: 'It's a Long Way to Furudu', *60 Minutes* (1979), retrieved May 2021 at https://www.youtube.com/watch?v=LGRmuFqel4E

CHAPTER 6

77 *The Solomon Islands were named . . .*: Lord Amherst of Hackney and B. Thomson (eds), *The Discovery of the Solomon Islands by Alvaro de Mendaña in 1568*, Bedford Press (1901).

77 *nearly a centimetre each year*: S. Albert et al., 'Interactions between sea-level rise and wave exposure on reef island dynamics in the Solomon Islands', *Environ. Res. Lett.* 11(5) (2016).

77 *changes in wind patterns*: M. Merrifield and M. Maltrud, 'Regional sea level trends due to a Pacific trade wind intensification', *Geophys. Res. Lett.* 38(21) (2011).

77 *linked to global warming*: C. Fang and L. Wu, 'The role of ocean dynamics in tropical Pacific SST response to warm climate in a fully coupled GCM', *Geophys. Res. Lett.* 35 (2008).

78 *inhabited for at least three millennia*: P. Sheppard and R. Walter, 'A Revised Model of Solomon Islands Culture History', *Journal of the Polynesian Society* 115(1), 47–76 (2006).

78 *1943 journal article*: Capell, A., 'Notes on the Islands of Choiseul and New Georgia, Solomon Islands', *Oceania* 14(1), 20–29 (1943).

78 *suppressed by the twin influences*: D. R. Lawrence, Chapter 8, *The Naturalist and his 'Beautiful Islands': Charles Morris Woodford in the Western Pacific*, ANU Press (2014).

80 *tambu, or taboo, is a word . . .*: 'Taboo.' *Encyclopaedia Britannica.* Retrieved from https://www.britannica.com/topic/taboo-sociology

83 *A 1944* New Yorker *story*: Hersey, J., 'Survival', *The New Yorker,* 10 June 1944.

84 *The gross domestic product of the whole country*: World Bank data retrieved from https://data.worldbank.org/country/solomon-islands

85 *signed a financing agreement*: 'GCF and World Bank kick off hydropower project in Solomon Islands', Green Climate Fund press release (2 August 2019).

85 *falling short of their promise*: J. Shankleman, 'Rich Countries Missing the $100 Billion Climate Finance Goal', Bloomberg (6 November 2020).

85 *Of the cash so far allocated by the GCF . . .*: Green Climate Fund project portfolio. Retrieved from https://www.greenclimate.fund/projects/dashboard

86 *The dolphins drew the sea people*: D. Takekawa, 'Hunting method and the ecological knowledge of dolphins among the Fanalei

villagers of Malaita, Solomon Islands', *SPC Traditional Marine Resource Management and Knowledge Information Bulletin* 12, 3–11 (2000).

87 *over 1,600 kills*: The hunt resumed in 2013 after a three-year suspension. M. Oremus et al., 'Resumption of traditional drive hunting of dolphins in the Solomon Islands in 2013', *R. Soc. Open Sci.* 2:140524 (2015).

89 *Like most Malaitans, he's not even connected to an electric grid*: Three per cent of households in Malaita were connected to an electric grid, according to the 2009 national census. *2009 Population & Housing Census, National Report* (Vol. 2, p. xxi). Solomon Islands National Statistical Office (2009).

CHAPTER 7

93 *far more densely inhabited than any other large country*: World Bank data retrieved in May 2021 from https://data.worldbank.org/indicator/EN.POP.DNST; https://data.worldbank.org/indicator/AG.LND.TOTL.K2; https://data.worldbank.org/indicator/SP.POP.TOTL

94 *over two thousand square kilometres of Bangladesh*: M. K. Hasan et al., 'Inundation modelling for Bangladeshi coasts using downscaled and bias-corrected temperature', *Climate Risk Management* 27, 100207 (2020).

94 *Recent academic studies back up her suspicion*: T. K. Das, 'Determination of Drinking Water Quality: A Case Study on Saline Prone South-West Coastal Belt of Bangladesh', *Environ. Sci. & Natural Resources,* 10(1), 101–8 (2017); M. A. Rakib et al., 'Groundwater salinization and associated co-contamination risk increase severe drinking water vulnerabilities in the southwestern coast of Bangladesh', *Chemosphere* 246, 125646 (2020); D. K. Das, 'Health cost of salinity contamination in drinking water: evidence from Bangladesh', *Environmental Economics and Policy Studies* 21, 371–97 (2019).

95 *biggest export earner after clothing and textiles*: Data retrieved from website of Bangladesh Bank at https://www.bb.org.bd/econdata/export/exp_rcpt_comodity.php

95 *kill dozens of people each year*: C. Inskip et al., 'Human–Tiger Conflict in Context: Risks to Lives and Livelihoods in the Bangladesh Sundarbans', *Hum Ecol* 41, 169–86 (2013).

95 *according to World Bank predictions: Groundswell: Preparing for Internal Climate Migration.* World Bank report (March 2018).

97 *the average square kilometre of the capital: Population & Housing Census 2011.* Vol. 3: Urban Area Report, p. 44. Bangladesh Bureau of Statistics (2014).

97 *migrants have continued to pour into Dhaka*: M. Al Amin, 'Dhaka, Chittagong destination of 80% internal migrants', *Dhaka Tribune* (24 November 2018).

97 *Most of them heading for overcrowded slum districts*: M. Ahmed, 'Health and Well-Being of Climate Migrants in Slum Areas of Dhaka', in W. Leal Filho et al. (eds), *Good Health and Well-Being, Encyclopedia of the UN Sustainable Development Goals*, Springer (2020).

98 *the World Bank predicts: Groundswell: Preparing for Internal Climate Migration.* World Bank report (March 2018).

99 *one of several reporters targeted*: H. Cockburn, 'Bangladeshi journalist arrested and another on run after writing about voting irregularities in election', *Independent,* 2 January 2019.

99 *Reporters Without Borders' press freedom index*: Retrieved at https://rsf.org/en/ranking_table

99 *chair of the Climate Vulnerable Forum*: 'Bangladesh: Chair of Climate Vulnerable Nations' Forum', CVF press release (12 June 2020).

99 *UN Champion of the Earth award*: 'Hasina receives Champions of the Earth award', *Daily Star,* 28 September 2015.

100 *China is providing nearly half the funding*: R. K. Byron, '$2.76b Chinese Fund: Quick move to execute 3 projects', *Daily Star*, 20 October 2016.

102 *the leader of a landless women's group was murdered*: The 1990 killing of Karunamayee Sarder, as described in M. Guhathakurta, 'Globalization, Class and Gender Relations: The shrimp industry in southwestern Bangladesh', *Development* 51, 212–19 (2008).

CHAPTER 8

107 *'If Venice does not have . . .'*: Facebook post by Luigi Di Maio (translated from the Italian by the author). Retrieved from https://www.facebook.com/LuigiDiMaio/posts/2601632349873295

110 *the tide rose 187 centimetres*: Data provided by the Centro Previsioni e Segnalazioni Maree, Città di Venezia.

110 *The water surged into . . .*: 'Flooded Venice battles new tidal surge', BBC News (15 November 2019).

111 *According to a study published by ten European academics*: P. Lionello et al., 'Extremes floods of Venice: characteristics, dynamics, past and future evolution', *Nat. Hazards Earth Syst. Sci. Discuss.* [preprint] (2020).

111 *the creation of the city itself*: T. Madden, *Venice: A New History*, Penguin (2012), p.14.

111 *Then, during the Renaissance . . .*: P. Furlanetto, and A. Bondesan, 'Geomorphological evolution of the plain between the Livenza and Piave Rivers in the sixteenth and seventeenth centuries inferred by historical maps analysis (Mainland of Venice, Northeastern Italy)', *Journal of Maps* 11(2), 261–6 (2015).

113 *presiding over a feeding frenzy*: 'Venezia, decine di milioni di tangenti e una ventina di condanne: perché non c'è il Mose a proteggere la città', *Il Fatto Quotidiano*, 13 November 2019.

114 *The police swooped*: 'Mose: arrestati sindaco Venezia e assessore Veneto', *Altalex*, 4 June 2014.

114 *the worst of any major Western economy*: Transparency International Corruption Perceptions Index retrieved at https://www.transparency.org/en/cpi/2020

117 *threw central London into gridlock*: M. Taylor and D. Gayle, 'Dozens arrested after climate protest blocks five London bridges', *Guardian*, 17 November 2018.

117 'We . . . declare ourselves in rebellion . . .': Extinction Rebellion, *This Is Not A Drill: An Extinction Rebellion Handbook*, Penguin (2019).

CHAPTER 9

125 *has overseen the street shootings of several thousand alleged felons*: See, for example, Amnesty International, *Extrajudicial Executions In The Philippines' 'War On Drugs'* (2017).

126 *a peak force of nearly 200 miles per hour*: Joint Typhoon Warning Center. *Annual Tropical Cyclone Report 2013*. Retrieved from https://www.metoc.navy.mil/jtwc/products/atcr/2013atcr.pdf

127 *how Haiyan's force had been amplified*: For example, Takayabu, I., 'Climate change effects on the worst-case storm surge: a case study of Typhoon Haiyan', *Environ. Res. Lett.* 10, 064011 (2015).

128 *In the wake of his brutal attempt . . .*: S. Mydans, 'Marcos Flees

And Is Taken To Guam; U.S. Recognizes Aquino As President', *New York Times,* 26 February 1986.

128 *a petition from local activists*: Petition to the Commission on Human Rights of the Philippines, submitted by Greenpeace Southeast Asia and Philippine Rural Reconstruction Movement. Retrieved from https://www.greenpeace.org/static/planet4-philippines-stateless/2019/05/cd8c5ca1-cd8c5ca1-human_rights_and_climate_change_consolidated_reply_2_10_17.pdf

129 *twenty-one children had sued the national government*: R. Salas et al., 'The Case of Juliana v. U.S. — Children and the Health Burdens of Climate Change', *N Engl J Med* 380, 2085–7 (2019).

129 *Saúl Luciano Lliuya was launching legal action*: D. Collins, 'Peruvian farmer demands climate compensation from German company', *Guardian,* 16 March 2015.

130 *a groundbreaking address to the US Senate*: P. Shabecoff, 'Global Warming Has Begun, Expert Tells Senate', *New York Times,* 24 June 1988.

131 *Shell paid out more money to its shareholders . . .*: Data retrieved from S&P Capital IQ database.

131 *Ben van Beurden earned . . .*: 'Shell CEO's pay more than doubles to $22.8 million in 2018', Reuters (14 March 2019).

133 *the single biggest customer for the global jewellery industry*: S. Seagrave, *The Marcos Dynasty,* Lume Books (2017), p. 249.

134 *celebrated by media*: N. Dizon, 'Leyte lantern reminds survivors they can rise again', *Philippine Daily Inquirer,* 24 December 2013.

137 *disaster capitalism*: N. Klein, *The Shock Doctrine: The Rise of Disaster Capitalism,* Penguin (2014).

CHAPTER 10

141 *its annual turnover is about $70 billion*: Data retrieved from S&P Capital IQ database.

141 *the biggest beast in reinsurance*: H. Rupawala, 'Munich Re defends top global reinsurance ranking in 2019', S&P Global Market Intelligence (2 June 2020); J. F. Outreville, 'The World's Largest Reinsurance Groups: A look at names, numbers and countries from 1980 to 2010', *Insurance and Risk Management* 80(1), 137–56 (2012).

142 *when news leaked of an orgy*: S. Evans, 'German insurer Munich Re held orgy for salesmen', BBC News (20 May 2011)

142 *Hurricane Andrew smashed . . .*: D. Victor, 'Hurricane Andrew: How The Times Reported the Destruction of 199', *New York Times*, 6 September 2017.

143 *Sixteen insurance groups went bankrupt*: Insurance Information Institute. *Hurricane Andrew Fact Sheet.* Retrieved from https://www.iii.org/article/hurricane-andrew-fact-sheet#_msocom_2

143 *a record five years running*: National Centers for Environmental Information. '2020 North Atlantic Hurricane Season Shatters Records.' Retrieved from https://www.ncei.noaa.gov/news/2020-north-atlantic-hurricane-season-shatters-records

143 *In 2017 alone*: National Hurricane Center Tropical Cyclone reports. Retrieved from https://www.nhc.noaa.gov/data/tcr/AL092017_Harvey.pdf; https://www.nhc.noaa.gov/data/tcr/AL112017_Irma.pdf; https://www.nhc.noaa.gov/data/tcr/AL152017_Maria.pdf

144 *with a total damage bill*: Total damage costs in current US dollars, as estimated by the National Hurricane Center. 'Costliest U.S. tropical cyclones tables updated.' NHC report retrieved from https://www.nhc.noaa.gov/news/UpdatedCostliest.pdf

144 *His account of the event*: W. Strachey, *A True Reportory of the Wreck.* Retrieved from the Virtual Jamestown website: http://www.virtualjamestown.org/fhaccounts_date.html#1600

145 *64,000 people*: World Bank data retrieved from https://data.worldbank.org/indicator/SP.POP.TOTL

145 *Nearly $100 billion*: Aon Securities. *ILS Annual Report 2020.* Retrieved fromhttp://thoughtleadership.aon.com/documents/280920_aon_securities_ils_annual_2020_update.pdf

145 *If you're a Coca-Cola employee . . .*: Artemis data retrieved in May 2021 from https://www.artemis.bm/pension-funds-investing-in-insurance-linked-securities-ils/

146 *the world's biggest ILS fund manager*: Artemis data retrieved in May 2021 from https://www.artemis.bm/ils-fund-managers/

147 *the value of catastrophe bonds held firm*: S. Hills, 'Swiss Re cat bond index finishes 2009 on all-time high', Reuters (14 January 2010).

146 *pension funds holding a total of $32 trillion*: 'Pension Funds in Figures.' Report published by the OECD (June 2020).

148 *worst wildfires in its history*: California Department of Forestry and Fire Protection. '2020 Fire Season.' Retrieved from https://www.fire.ca.gov/incidents/2020/

148 *unprecedented outbreak of forest blazes in Australia*: A. Filkov et al. 'Impact of Australia's catastrophic 2019/20 bushfire season on

communities and environment. Retrospective analysis and current trends', *Journal of Safety Science and Resilience* 1(1), 44–56 (2020).

148 *thirty major storms*: 'Record-breaking Atlantic hurricane season draws to an end.' Report by NOAA (24 November 2020). Retrieved from https://www.noaa.gov/media-release/record-breaking-atlantic-hurricane-season-draws-to-end

148 *Miguel's farm*: This name has been changed.

149 *killing 165 people and destroying property worth nearly $40 billion*: AON. *Global Catastrophe Recap* (November 2020). Retrieved from http://thoughtleadership.aon.com/Documents/20201210_analytics-if-november-global-recap.pdf

149 *only twice before*: According to the National Hurricane Center, Eta reached maximum sustained wind speeds of 130 knots (150mph). I ran an analysis of the NHC's HURDAT2 dataset that showed this speed had only been reached or exceeded by two prior hurricanes: an unnamed hurricane that reached 150 knots in 1932, and Hurricane Lenny, which reached 135 knots in 1999. Sources: National Hurricane Center. 'Hurricane Eta Discussion Number 10.' Retrieved from https://www.nhc.noaa.gov/archive/2020/al29/al292020.discus.010.shtml?; NHC HURDAT2 dataset retrieved from https://www.aoml.noaa.gov/hrd/hurdat/hurdat2.html

149 *just 15 miles south*: National Hurricane Center. 'Hurricane Iota Tropical Cyclone Update.' Retrieved from https://www.nhc.noaa.gov/archive/2020/al31/al312020.update.11170347.shtml?

149 *Between them the two hurricanes . . .*: AON. *Global Catastrophe Recap* (November 2020). Retrieved from http://thoughtleadership.aon.com/Documents/20201210_analytics-if-november-global-recap.pdf; Kitroeff, N., '2 Hurricanes Devastated Central America. Will the Ruin Spur a Migration Wave?' *New York Times*, 4 December 2020; M. Villareal, 'Tropical Storm Eta hits North Carolina with deadly flooding', CBS News (13 November 2020).

150 *a dozen people who died*: J. Bow, 'Confirman muerte de 12 personas por deslave en Macizo de Peñas Blancas', *Confidencial*, 18 November 2020.

150 *Central America's largest and poorest country*: World Bank data retrieved from https://data.worldbank.org/indicator/NY.GDP.PCAP.CD; https://data.worldbank.org/indicator/AG.LND.TOTL.K2.

151 *one of the world's twenty most corrupt nations*: Transparency International Corruption Perceptions Index, retrieved at https://www.transparency.org/en/cpi/2019

151 *The Atlantic hurricane season of 2017 had been . . .*: E. Faust and M. Bove, 'The hurricane season 2017: a cluster of extreme storms', report published by Munich Re (1 December 2017).
151 *XL, meanwhile, has been getting its own protection . . .*: S. Evans, 'AXA XL settles for $475m Galileo Re 2019-1 catastrophe bond', Artemis (9 December 2019).

CHAPTER 11

159 *lowest global production since 1957*: 'World wine output falls to 60-year low', Reuters (24 April 2018).
159 *the worst extended drought on national record*: R. Garreaud et al., 'The Central Chile Mega Drought (2010–2018): A climate dynamics perspective', *International Journal of Climatology* 40(1), 421–39 (2020).
159 *the biggest wildfires in Chile's recorded history*: H. Dacre et al., 'Chilean Wildfires: Probabilistic Prediction, Emergency Response, and Public Communication', *Bulletin of the American Meteorological Society* 99(11), 2259–74 (2018).
160 *42,000 tonnes of water*: J. Cifuentes, 'Ya está en Chile el 'Supertanker' ruso', *La Tercera*, 30 January 2017.
160 *the Chinchorro people began mummifying their dead*: 'Chile seeks help to protect world's oldest mummies', Reuters (27 October 2016).
161 *Augusto Pinochet's secret police in the 1970s*: 'Chilean Mass Grave Believed to Contain 26 Bodies of Leftists', Associated Press (25 July 1990).
161 *the closest thing on earth to the Martian surface*: 'Mars Rover Tests Driving, Drilling and Detecting Life in Chile's High Desert', NASA press release (13 March 2017).
161 *a desert far older and drier than the Sahara*: J. Houston, 'Variability Of Precipitation In The Atacama Desert: Its Causes And Hydrological Impact', *Int. J. Climatol.* 26: 2181–98 (2006); O. Kelley, 'Where the Least Rainfall Occurs in the Sahara Desert, the TRMM Radar Reveals a Different Pattern of Rainfall Each Season', *Journal of Climate* 27(18), 6919–39 (2014); H. Le Houérou, 'Outline of the biological history of the Sahara', *Journal of Arid Environments* 22(1), 3–30 (1992); J. Clarke, 'Antiquity of aridity in the Chilean Atacama Desert', *Geomorphology* 73, 101–14 (2006).
161 *rainfall over Cerro Honar set to decline*: J. Marengo et al., 'Climate

Change: Evidence and Future Scenarios for the Andean Region', pp. 110–27, in S. Herzog et al (eds), *Climate Change and Biodiversity in the Tropical Andes*, Inter-American Institute for Global Change Research and Scientific Committee on Problems of the Environment (2011).

162 *according to communities to the south*: D. Sherwood, 'Indigenous groups in Chile's Atacama push to shut down top lithium miner SQM', Reuters (14 August 2020).

163 *having surfed the wave of export-driven Chilean growth*: For an in-depth history of the Chilean wine industry, see J. Del Pozo, *Historia del Vino Chileno*, LOM Ediciones (2014).

163 *now worth nearly $2 billion a year*: 2019 export data retrieved from UN Comtrade database.

164 *some of the country's top climate scientists wrote*: R. Garreaud et al., 'The Central Chile Mega Drought (2010–2018): A climate dynamics perspective', *International Journal of Climatology* 40(1), 421–39 (2020).

164 *just 82 millimetres of rain*: Data retrieved from the website of the Dirección Meteorológica de Chile: https://climatologia.meteochile.gob.cl/application/anual/aguaCaidaAnual/330020/2019

164 *lower than the average figure for famously arid Dubai*: Data for Dubai retrieved from the website of the World Meteorological Organisation: http://worldweather.wmo.int/en/city.html?cityId=1190

166 *gradually easing the irrigation rules*: F. Martin, 'The Irrigation of grapevines in Europe – an update on existing legislation', *Irrigazette*, 28 October 2016; A. Mileham, 'INAO to allow AOCs to trial climate-change varieties', *The Drinks Business*, 15 November 2018.

167 *gathering two hundred of them for a dinner*: E. Moya, 'Spanish winemaker Torres warms to environmentalism', *Guardian*, 29 October 2009.

167 *have started winning prizes*: R. Smithers, 'English wines win record number of awards in global tasting competition', *Guardian*, 22 September 2020.

167 *top French champagne marques buying up land*: C. Tominey, 'Santé! Another Champagne giant uncorks plans to buy an English vineyard', *The Telegraph*, 25 July 2020.

167 *including one near a shrinking glacier*: B. Pancevski, 'Chateau Viking: Climate Change Makes Northern Wine a Reality', *Wall Street Journal*, 29 October 2019.

CHAPTER 12

173 *the hottest inhabited place on earth*: M. Fazzini, 'The Climate of Ethiopia', in P. Billi (ed.,) *Landscapes and Landforms of Ethiopia. World Geomorphological Landscapes*, Springer (2015).

173 *perhaps a single day of modest rainfall each month*: Modelled climate data for Semera, capital of Afar region, retrieved from Meteoblue website: https://www.meteoblue.com/en/weather/historyclimate/climatemodelled/semera_ethiopia_6913519

175 *including insect pests*: P. Lehmann et al., 'Complex responses of global insect pests to climate warming', *Frontiers in Ecology and the Environment* 18(3), 141–50 (2020).

175 *an exceptionally intense pair of cyclones*: A. Salih et al., 'Climate change and locust outbreak in East Africa', *Nature Climate Change* 10, 584–5 (2020).

175 *linked, studies suggest, to rising sea surface temperatures*: C. Sun et al., 'Recent Acceleration of Arabian Sea Warming Induced by the Atlantic-Western Pacific Trans-basin Multidecadal Variability', *Geophysical Research Letters* 46(3), 1662–71 (2019).

175 *each of them eating its own body weight*: Food and Agriculture Organization: Desert Locust Information Service of the Migratory Pests Group. 'Frequently Asked Questions.' Retrieved from: http://www.fao.org/ag/locusts/oldsite/LOCFAQ.htm

176 *41 per cent of Afari children who are physically stunted*: USAID. 'Ethiopia: Nutrition Profile.' Retrieved from: https://www.usaid.gov/sites/default/files/documents/1864/Ethiopia-Nutrition-Profile-Mar2018-508.pdf

177 *in the highland areas that fed the Shebelle . . .*: S. Rosell, 'Regional perspective on rainfall change and variability in the central highlands of Ethiopia, 1978–2007', *Applied Geography* 31, 329–38 (2011).

178 *the brutal tactics of Abdi Illay*: Human Rights Watch, *'We Are Like The Dead': Torture and other Human Rights Abuses in Jail Ogaden, Somali Regional State, Ethiopia.* Report published July 2018.

180 *over a million people displaced*: 'Over 1 Million People Displaced due to Conflict in Northern Ethiopia: IOM DTM', IOM press release (23 April 2021).

180 *extensive reports of mass rape and massacres*: For example, D. Patta, 'Reports of executions and mass-rape emerge from the obscured war in Ethiopia's Tigray region', CBS News (25 March 2021).

180 *350 million saplings in twelve hours*: S. Paget and H. Regan, 'Ethiopia plants more than 350 million trees in 12 hours', CNN (30 July 2019).

180 *wants that number to reach twenty billion*: J. Myers, 'Ethiopia wants to plant 5 billion seedlings this year', World Economic Forum (5 June 2020).

180 *as much as 40 per cent*: B. Bishaw, 'Deforestation and Land Degradation in the Ethiopian Highlands: A Strategy for Physical Recovery', *Northeast African Studies* 8(1), 7–25 (2001).

181 *a factor behind the devastating famine*: T. Vestal, 'Famine in Ethiopia: Crisis of Many Dimensions', *Africa Today* 32(4), 7–28 (1985).

181 *climate change has been dragging down average rainfall*: I. Niang et al., 'Africa', in *Climate Change 2014: Impacts, Adaptation, and Vulnerability. Part B: Regional Aspects. Contribution of Working Group II to the Fifth Assessment Report of the Intergovernmental Panel on Climate Change* (2014), p. 1209.

182 *Nine-tenths of its electricity*: D. Conway et al., 'Hydropower plans in eastern and southern Africa increase risk of concurrent climate-related electricity supply disruption', *Nature Energy* 2, 946–53 (2017).

182 *Ethiopia's total electricity capacity*: M. Tafesse et al., 'Electricity regulation in Ethiopia: overview.' Retrieved from Thomson Reuters Practical Law website at https://uk.practicallaw.thomsonreuters.com/w-028-1702; UK data retrieved from https://www.gov.uk/government/collections/electricity-statistics

182 *rising temperatures predicted to drive up its farmers water needs*: H. Eid, 'Assessing the Economic Impacts of Climate Change on Agriculture in Egypt', World Bank policy research working paper (July 2007).

183 *Nationalist media pundits in Cairo . . .*: S. Magdy, 'Egyptian media urges military action against Ethiopia as Nile talks break down', Associated Press (22 October 2019).

CHAPTER 13

187 *the annual volume of rain during the monsoon season . . .*: India Meteorological Department, 'Observed Rainfall Variability and Changes over Maharashtra State.' Retrieved from https://imdpune.gov.in/hydrology/rainfall%20variability%20page/maharashtra_final.pdf

188 *another 3,927*: 'Farmers' suicides highest in Maharashtra despite loan waiver, reform measures', *The Indian Express* (11 October 2020).

188 *The world's population is set to reach nearly ten billion by 2050*: United Nations *World Population Prospects 2019*. Retrieved from https://population.un.org/wpp/.

188 *global food demand will rise even more quickly*: E. Fukase, and W. Martin, 'Economic growth, convergence, and world food demand and supply', *World Development* 132, 104954 (2020).

189 *a new US study will reach a shocking conclusion*: D. Ray et al., 'Climate change has likely already affected global food production.', *PLoS ONE* 14(5): e0217148 (2019).

189 *William and Paul Paddock wrote*: W. Paddock and P. Paddock, *Famine 1975! America's Decision: Who Will Survive?*, Little, Brown (1967).

190 *nearly fifteen million more mouths to feed*: World Bank data retrieved from https://data.worldbank.org/indicator/SP.POP.GROW

191 *equipped to resist soaring temperatures*: A. Grover et al., 'Generating high temperature tolerant transgenic plants: Achievements and challenges', *Plant Sci.* 2013 May (205–206), 38–47 (2013).

191 *the upswell of pests that they will bring*: C. Deutsch et al., 'Increase in crop losses to insect pests in a warming climate', *Science* 361(6405), 916–19.

192 *Vandana has become known as the 'rock star'*: 'Vandana Shiva on why the food we eat matters', BBC Travel (28 January 2021).

192 *Her fans include Prince Charles*: M. Specter, 'Seeds of Doubt', *The New Yorker*, 18 August 2014.

193 *Leading bodies such as the World Health Organisation*: World Health Organisation, 'Food, Genetically Modified'. Retrieved 11 May 2021 from https://www.who.int/health-topics/food-genetically-modified#tab=tab_2

194 *roundly dismissed by leading scientists*: Most notably a 2014 study on GM maize that was retracted by the publishing journal after a wave of criticism from scientists in the field: G. Seralini et al., [RETRACTED] 'Long term toxicity of a Roundup herbicide and a Roundup-tolerant genetically modified maize', *Food and Chemical Toxicology* 50(11), 4221–31 (2012).

194 *most notably in Europe*: The Royal Society, 'What GM crops are currently being grown and where?'. Retrieved from https://

royalsociety.org/topics-policy/projects/gm-plants/what-gm-crops-are-currently-being-grown-and-where/

194 *Scientists in countries like Tanzania and Kenya*: M. Lynas, *Seeds of Science*, Bloomsbury Sigma (2018); H. Heuler, 'In Kenya, Calls Grow to Lift Controversial GMO Ban', VOA (20 November 2014).

194 *concerns about contamination*: D. Wafula and G. Gruère, 'Genetically Modified Organisms, Exports, and Regional Integration in Africa', in J. Falck-Zepeda et al., Genetically Modified Crops in Africa, IFPRI (2013).

CHAPTER 14

201 *Uvs, Mongolia's northwesternmost province . . .*: Uvs data retrieved from the provincial government statistical website at http://uvs.nso.mn/

204 *the frequency and severity of dzud events has been increasing*: M. Rao et al., 'Dzuds, droughts, and livestock mortality in Mongolia', *Environmental Research Letters* 10(7), 074012 (2015); B. Nandintsetseg, 'Cold-season disasters on the Eurasian steppes: Climate-driven or man-made', *Sci Rep.* 8, 14769 (2018); 'П.Гомболүүдэв: Байгалийн горим алдагдаж ган зудын давтамж нэмэгдэж байна', GoGo Mongolia (30 May 2019).

204 *At the start of this century . . .*: Rao et al. (2015).

204 *consistent with scientists' warnings*: Y. Ijima and M. Hori, 'Cold air formation and advection over Eurasia during "dzud" cold disaster winters in Mongolia', *Natural Hazards* 92, 45–56 (2018); J. Liu et al., 'Impact of declining Arctic sea ice on winter snowfall', *PNAS* 109(11), 4074–9 (2012); M. Mori et al., 'Robust Arctic sea-ice influence on the frequent Eurasian cold winters in past decades', *Nature Geoscience* 7, 869–73 (2014); M. Mori et al., 'A reconciled estimate of the influence of Arctic sea-ice loss on recent Eurasian cooling', *Nature Climate Change* 9, 123–9 (January 2019).

204 *the volume lost each year*: A. Kumar et al., 'Global warming leading to alarming recession of the Arctic sea-ice cover: Insights from remote sensing observations and model reanalysis', Heliyon 6(7), e04355 (2020).

204 *Winter snowfall in Mongolia . . .*: Third National Communication of Mongolia under the UNFCCC (2018).

204 *Similar factors may be driving . . .*: J. Francis & S. Vavrus, 'Evidence linking Arctic amplification to extreme weather in mid-latitudes,'

Geophysical Research Letters 39 (6) (2012); J. Cohen et al, *Arctic change and possible influence on mid-latitude climate and weather: A US CLIVAR white paper* (No. 2018-1). Washington, DC: U.S. CLIVAR Project Office (2018); O. Milman, 'Heating Arctic may be to blame for snowstorms in Texas, scientists argue', *The Guardian* (17 February 2021).

204 *average temperature in Mongolia up more than 2°*: Ibid.

206 *with herders given production targets set out in five-year plans*: R. Mearns, 'Pastoralists, Patch Ecology and Perestroika: Understanding Potentials for Change in Mongolia', *IDS Bulletin* 22(4), 25–33 (1991).

206 *has surged to seventy million*: T. Baljmaa, 'Mongolia has 70.9 million livestock animals counted', Montsame (27 December 2019).

206 *triple the number in the communist* era: Erdenesan Eldevochir, Mongolian National Registration and Statistics Office, 'Livestock Statistics in Mongolia'. Retrieved from http://www.fao.org/fileadmin/templates/ess/documents/apcas26/presentations/APCAS-16-6.3.5_-_Mongolia_-_Livestock_Statistics_in_Mongolia.pdf

207 *a quarter of people raise livestock*: 233,300 of 897,427 in 2019, according to government data. Sources: Mongolian Ministry of Agriculture, 'Цаг үеийн мэдээлэл дугаар 2019/12'. Retrieved from https://mofa.gov.mn/exp/ckfinder/userfiles/files/ts201912.pdf; Aggregate national household data retrieved from Mongolian General Statistical Database at https://www.1212.mn/tables.aspx?TBL_ID=DT_NSO_0300_033V1

CHAPTER 15

213 *It is a fallacy to say*: Author's translation of the original Portuguese, retrieved from the website of Brazil's Câmara dos Deputados: https://www.camara.leg.br/noticias/589803-bolsonaro-amazonia-nao-e-patrimonio-da-humanidade-nem-pulmao-do-mundo/

215 *it vanished the night they stabbed Ari in the neck*: Having inspected Ari's body, Awapy rejects local authorities' conclusion that he was killed with a blunt object.

215 *Two-thirds of the tribe's 250 people*: A. De Almeida Silva, *Entre a floresta e o concreto: Os impactos socioculturais no povo indígena Jupaú em Rondônia*, Paco e Littera (2015).

215 *second-loudest bird*: J. Podos and M. Cohn-Haft, *Current Biology* 29(20), PR1068–R1069 (2019).

216 *1.9 million hectares*: Conselho Nacional de Direitos Humanos, *Relatório: Missão de Levantamento de Informações Sobre a Terra Indígena Uru Eu Wau Wau*. Retrieved from https://www.gov.br/participamaisbrasil/blob/baixar/3309

216 *the promise of legal rights*: H. Binswanger, 'Brazilian policies that encourage deforestation in the Amazon', *World Development* 19(7), 821–9 (1991).

216 *over 300,000 square miles of it have been lost*: D. Conrado da Cruz et al., 'An overview of forest loss and restoration in the Brazilian Amazon', *New Forests* 52, 1–16 (2021).

216 *most of this land turned into cattle pasture*: R. Carvalho et al., 'Diversity of cattle raising systems and its effects over forest regrowth in a core region of cattle production in the Brazilian Amazon', *Regional Environmental Change* 20, Article 44 (2020).

216 *the biggest beef exporter in the world*: W. Alcântara da Silva Neto and M. Rumenos Piedade Bacchi, 'Growth of Brazilian beef production: effect of shocks of supply and demand', *Rev. Econ. Sociol. Rural* 52(2) (2014).

216 *broader slowdown of deforestation*: 'Brazil: Amazon deforestation falls to new low', BBC News (1 December 2010).

217 *grotesque bullying of a female lawmaker*: J. Anderson, 'Jair Bolsonaro's Southern Strategy', *The New Yorker*, 25 March 2019.

217 '*the Brazilian cavalry . . .*': Retrieved from the official gazette of Brazil's Câmara dos Deputados for 16 April 1998 at http://imagem.camara.gov.br/Imagem/d/pdf/DCD16ABR1998.pdf#page=33

219 *the death of Rieli Franciscato*: 'Amazon expert killed with arrow while working to protect uncontacted tribes in Brazil', CNN (11 September 2020).

220 *a new world record of 1.8 million tonnes*: 'Exportação de carne bovina brasileira bateu recorde em 2019', Canal Rural (6 January 2020).

222 *removing more than 400 million tonnes of carbon dioxide*: O. Phillips et al., 'Carbon uptake by mature Amazon forests has mitigated Amazon nations' carbon emissions', *Carbon Balance and Management 12*, Article 1 (2017).

222 *the annual emissions of the UK or Australia*: 2018 emissions data retrieved from the website of the Union of Concerned Scientists: https://www.ucsusa.org/resources/each-countrys-share-co2-emissions

223 *Since the 1970s, scientists have understood . . .*: E. Salati et al.,

'Recycling of water in the Amazon Basin: An isotopic study', *Water Resources Research* 15(5), 1250–58 (1979).

223 *the tipping point is likely to be reached . . .*: T. Lovejoy and C. Nobre, 'Amazon Tipping Point', *Science Advances* 4(2), eaat2340 (2018).

223 *above 17 per cent*: Carlos Nobre's estimate. Some other estimates are slightly higher.

224 *a lengthening dry season*: J. Agudelo et al., 'Influence of longer dry seasons in the Southern Amazon on patterns of water vapor transport over northern South America and the Caribbean', *Climate Dynamics* 52, 2647–65 (2019).

224 *declining rates of carbon absorption*: R. Brienen et al., 'Long-term decline of the Amazon carbon sink', *Nature* 519, 344–8 (2015).

224 *rising mortality among the tree species . . .*: A. Esquivel-Muelbert et al., 'Tree mode of death and mortality risk factors across Amazon forests', *Nature Communications* 11, Article 5515 (2020).

224 *'We are victims . . .'*: Author's translation of Jair Bolsonaro's remarks to the UN General Assembly on 22 September 2020, retrieved from https://funag.gov.br/index.php/pt-br/2015-02-12-19-38-42/3334

225 *prosecutors who accuse him . . .*: Prosecutors' testimony to the Judicial Section of Amazonas State, submitted 27 May 2019. Trial number 5253-29.2017.4.01.3200 (Medida Cautelar). Several reports on the case have been published by Amazonas Atual, for example: 'PMs atuavam como milícia para fazendeiros no Amazonas, diz MPF em relatório da Operação Ojuara' (1 March 2020).

227 *annual revenue of $51 billion*: Data for 2020 calendar year retrieved from S&P Capital IQ database.

228 *their holding company admitted . . .*: A. Schipani, 'Batista-family group in $3bn JBS plea deal', *Financial Times*, 31 May 2017.

CHAPTER 16

240 *legislation in some states stipulating . . .*: J. Bromwich and S. Yar, 'The Fake Meat War', *New York Times*, 25 July 2019.

240 *China imports three million tonnes of beef a year*: Data retrieved from UN Comtrade database.

242 *In 2015,* Consumer Reports *magazine sampled . . .*: A. Rock, 'How Safe Is Your Ground Beef?' *Consumer Reports*, 21 December 2015.

243 *about 15 million square miles is used to support animals:* This

includes both pasture and land used to grow animal feed: 40 million square kilometres, according to Our World In Data calculations using data from the Food and Agriculture Organisation. Retrieved from https://ourworldindata.org/global-land-for-agriculture

244 *But while CO_2 can linger . . .*: US Environmental Protection Agency, 'Understanding Global Warming Potentials'. Retrieved from https://www.epa.gov/ghgemissions/understanding-global-warming-potentials

244 *$9 million in startup funding*: Data retrieved from Crunchbase database.

245 *Glenn Beck marvelled on his radio show*: Retrieved from: https://www.glennbeck.com/radio/impossible-burger-blind-taste-test-can-meat-eaters-taste-the-fake-meat

245 *more than 30,000 restaurants and nearly 20,000 food shops*: Communication from Impossible Foods spokesperson.

CHAPTER 17

251 *'I think by 2020, if oil stops we can survive . . .'*: S. Nakhoul et al., 'Saudi prince unveils sweeping plans to end 'addiction' to oil', Reuters (25 April 2016).

253 *When the prince's grandfather . . .*: D. Rundell, *Vision or Mirage: Saudi Arabia at the Crossroads*, I.B. Tauris (2020), p.195

254 *many analysts say it will soon enter a steady decline*: See, for example, McKinsey & Company's Global Energy Perspective 2021. Retrieved from McKinsey's website: https://www.mckinsey.com/~/media/McKinsey/Industries/Oil%20and%20Gas/Our%20Insights/Global%20Energy%20Perspective%202021/Global-Energy-Perspective-2021-final.pdf

254 *MBS stamped his mark . . .*: S. Kalin and K. Paul, 'Saudi Arabia says it has seized over $100 billion in corruption purge', Reuters (30 January 2018).

254 *whom MBS has named as a role model*: P. Waldman, 'The $2 Trillion Project to Get Saudi Arabian Economy Off Oil', Bloomberg (21 April 2016).

255 *MBS put out a video*: 'Launch Announcement of THE LINE.' Retrieved at https://www.youtube.com/watch?v=xLN2Vu1iels

255 *MBS's imagination was fired . . .*: J. Scheck et al., 'A Prince's $500 Billion Desert Dream', *Wall Street Journal*, 25 July 2019; B. Hope

and J. Scheck, *Blood and Oil: Mohammed bin Salman's Ruthless Quest for Global Power*, John Murray (2020), pp.122–129; B. Hubbard, *MBS: The Rise to Power of Mohammed Bin Salman*, William Collins (2020), pp.168–170.

256 *It's in the sunniest region*: Data Retrieved from the World Bank's Global Solar Atlas at https://globalsolaratlas.info/

256 *strong winds*: A. B. Awana, 'Performance analysis and optimization of a hybrid renewable energy system for sustainable NEOM city in Saudi Arabia', *Journal of Renewable and Sustainable Energy* 11, 025905 (2019).

256 *dome-shaped, next-generation desalination plants*: V. Nereim and A. Di Paola, 'Saudis Plan "Solar Dome" Desalination Plants at Neom Mega-City', Bloomberg (29 January 2020).

257 *three-quarters of people are under forty*: 'Population In Saudi Arabia by Gender, Age, Nationality (Saudi / Non-Saudi) – Mid 2016 A.D.' Retrieved from the website of the General Authority for Statistics, Kingdom of Saudi Arabia: https://www.stats.gov.sa/en/5305

257 *sweeping away many of the country's most infamous strictures*: Hubbard, *MBS*.

257 *murder and dismemberment*: D. Kirkpatrick and C. Gall, 'Audio Offers Gruesome Details of Jamal Khashoggi Killing, Turkish Official Says', *New York Times*, 17 October 2018.

257 *Ernest Moniz*: T. Gardner, 'Ex-U.S. energy head Moniz halts Saudi work over journalist's disappearance', Reuters (10 October 2018).

257 *Neelie Kroes*: R. Jones and M. Stancati, 'Saudi Journalist's Disappearance Sends Chill Through Foreign Investors, Firms', *Wall Street Journal*, 11 October 2018.

257 *women who had campaigned for that move*: 'Saudi Arabia: Women's Rights Advocates Arrested.' Press release by Human Rights Watch (18 May 2018).

257 *rate of executions has surged*: 'Death penalty 2019: Saudi Arabia executed record number of people last year amid decline in global executions.' Press release by Amnesty International (21 April 2020).

257 *20,000 members of the Huwaitat tribe . . .*: J. Malsin and S. Said, 'Saudi Residents Push Back Against Crown Prince's Megacity Plan', *Wall Street* Journal, 20 April 2020.

258 *in the spring of 1938*: D. Yergin, *The Prize: The Epic Quest for Oil, Money & Power*, Simon & Schuster (2012), p.283.

258 *Aramco pumps . . .*: Data calculated by the author from the

companies' respective annual reports for 2020. Aramco's oil-equivalent production for the year was 12.4 million barrels/day. The combined production of ExxonMobil, Chevron, BP and Shell was 12.27 million barrels/day.

259 *market capitalization of two trillion dollars*: W. Kennedy and M. Martin, 'Prince Gets His $2 Trillion Aramco, But Victory Comes at a Cost', Bloomberg (12 December 2019).

259 *plastics and petrochemicals that account for . . .*: 'The Future of Petrochemicals.' Report from the International Energy Agency (October 2018). Available at https://www.iea.org/reports/the-future-of-petrochemicals

259 *up to 40 per cent of it*: M. Masnadi, 'Global carbon intensity of crude oil production', *Science* 361(6405), 851–3 (2018).

259 *Saudi Arabia already has the lowest emissions*: Ibid.

260 *JERA, the country's biggest electricity company . . .*: E. Yep and J. H. Tan, 'Japan's largest power producer JERA plans net zero CO2 by 2050', S&P Global Platts (13 October 2020).

260 *The EU, too, is targeting a massive expansion*: 'A hydrogen strategy for a climate-neutral Europe', Communication paper from the European Commission (8 July 2020).

261 *one of the world's biggest 'green hydrogen' plants*: C. Matthews and K. Blunt, 'Green hydrogen plant in Saudi desert aims to amp up clean power', *Wall Street Journal*, 8 February 2021.

CHAPTER 18

263 *These mines have got . . .*: D. Crowe, 'Scott Morrison backs coal wealth for decades to come', *Sydney Morning Herald*, 20 January 2021.

265 *the Brisbane Federal Court delivered*: N. McElroy, 'Bankrupted traditional owner vows to keep fighting Adani', *Brisbane Times*, 16 August 2019.

265 *his editor quietly cut a section*: R. Broome, *Aboriginal Australians: A History Since 1788*. Allen & Unwin (2010), p. 14.

266 *the global economic growth rate had been close to zero*: Angus Maddison estimated an average global economic growth rate of 0.05 per cent for the period 1500–1820. See A. Maddison, 'Measuring and Interpreting World Economic Performance 1500–2001', *Review of Income and Wealth* 51(1), 1–35 (2005).

266 *more than a third of the world's electricity generation*: 'Coal.'

Retrieved from the website of the International Energy Agency: https://www.iea.org/fuels-and-technologies/coal

266 *more coal than any other country*: Data Retrieved from UN Comtrade Database: https://comtrade.un.org/data/

266 *mining accounts for more than a tenth*: 'Composition of the Australian Economy.' Retrieved from the website of the Reserve Bank of Australia: https://www.rba.gov.au/snapshots/economy-composition-snapshot/

267 *incinerating 40,000 square miles of land*: 'Forest fire area data for the 2019–20 summer bushfire season in southern and eastern Australia.' Retrieved from the website of the Australian Government Department of Agriculture, Water and the Environment: https://www.agriculture.gov.au/abares/forestsaustralia/forest-data-maps-and-tools/fire-data#fire-area-and-area-of-forest-in-fire-area-by-jurisdiction

267 *killing 33 people*: 'Interim observations.' Report by the Australian Royal Commission into National Natural Disaster Arrangements. Retrieved from the Commission's website at https://naturaldisaster.royalcommission.gov.au/system/files/2020-08/Interim%20Observations%20-31%20August%202020_0.pdf

268 *some 40,000 people*: '2016 Census QuickStats.' Retrieved from the website of the Australian Bureau of Statistics: https://quickstats.censusdata.abs.gov.au/census_services/getproduct/census/2016/quickstat/SED30048

268 *killed more than half of its coral since 1995*: A. Dietzel et al., 'Long-term shifts in the colony size structure of coral populations along the Great Barrier Reef', *Proc. R. Soc. B.* 287 (2020).

268 *Locals had been infuriated*: M. Daley, 'The Stop Adani convoy – what really happened in north Queensland', *The Fifth Estate* (3 May 2019).

268 *Some liberals in Australia were so outraged*: M. Truu, '"Quexit": Some Australians want to ditch Queensland after election result', SBS News (19 May 2019).

269 *as the pollsters forecast*: P. Cockburn, and B. Kontominas, 'Election 2019: How the polls got it so wrong in predicting a Labor victory', ABC News (19 May 2019).

269 *reduce the country's carbon emissions by well over a third*: Labor promised to reduce emissions 45 per cent from 2005 levels by 2030. This represented a 37 per cent cut from 2019 levels. See J. Gabbatiss, 'Australian election 2019: What the manifestos say on

energy and climate change', *Carbon Brief* (29 April 2019); 'Quarterly Update of Australia's National Greenhouse Gas Inventory: March 2019.' Retrieved from the website of the Australian Government Department of the Environment and Energy: https://www.environment.gov.au/system/files/resources/6686d48f-3f9c-448d-a1b7-7e410fe4f376/files/nggi-quarterly-update-mar-2019.pdf

269 *who had once brandished a lump of coal:* Video Retrieved at https://www.theguardian.com/global/video/2017/feb/09/scott-morrison-brings-a-chunk-of-coal-into-parliament-video

270 *Tens of millions of years ago*: C. M. Barton et al., 'Latrobe Valley, Victoria, Australia: A world class brown coal deposit', *International Journal of Coal Geology* 23(1-4), 193–213 (1993).

270 *generated 85 per cent of the electricity used in all Victoria*: 'Our Coal Our Future – Future opportunities for brown coal.' Report by the State of Victoria (2008).

271 *The next will follow in 2028*: A. Morton, 'Yallourn, one of Australia's last brown coal power stations, to close early in favour of giant battery', *Guardian*, 10 March 2021.

272 *will run for only a year*: J. Whittaker et al., 'Victoria's Latrobe Valley coal-to-hydrogen pilot project gets green light from EPA', ABC News (14 February 2019).

273 *Australia's largest company and the world's biggest mining group*: By market capitalization, as of 7 May 2021. Verified by the author using S&P Capital IQ database.

273 *more than $40 billion worth of commodities a year*: Data retrieved from S&P Capital IQ database.

273 *about 600 million tonnes*: 'Scope 3 Emissions Calculation Methodology 2019.' Retrieved from BHP's website: https://www.bhp.com/-/media/documents/investors/annual-reports/2019/bhpscope3emissionscalculationmethodology2019.pdf?la=en

273 *Andrew Mackenzie announced*: N. Hume, 'BHP to set targets for reducing customers' carbon emissions', *Financial Times*, 23 July 2019.

274 *sell its thermal coal mines*: N. Hume, 'BHP responds to investor pressure with thermal coal exit', *Financial Times*, 18 August 2020.

275 *saying it will now put climate considerations at the centre*: Larry Fink's 2020 letter to CEOs, Retrieved from BlackRock's website:

https://www.blackrock.com/corporate/investor-relations/2020-larry-fink-ceo-letter

275 *Norway's trillion-dollar national wealth fund*: R. Milne, 'Norway's oil fund sells out of Glencore, Anglo American and RWE', *Financial Times*, 13 May 2020.

275 *which once made up a quarter of the S&P 500 stock index's value*: D. Bianco, 'S&P 500 sector composition: More tech, less energy than ever before', DWS Americas CIO View (14 February 2020).

275 *have seen that figure fall below 3 per cent*: D. Chisholm, 'Q2 2021 sector scorecard.' Retrieved from website of Fidelity Asset Management: https://www.fidelity.com/viewpoints/investing-ideas/quarterly-sector-update

CHAPTER 19

279 *About 200 million years ago*: A. Schettino and E. Turco, 'Breakup of Pangaea and plate kinematics of the central Atlantic and Atlas regions', *Geophys. J. Int.* 178, 1078–97 (2009).

279 *an extraordinary concentration of seismic activity*: P. Einarsson and B. Brandsdóttir, 'Seismicity of the Northern Volcanic Zone of Iceland', *Frontiers in Earth Science* 9, 166 (2021).

279 *smaller than Guatemala*: World Bank data Retrieved from https://data.worldbank.org/indicator/AG.LND.TOTL.K2

279 *a quarter of the population*: H. Sigurdsson, 'Volcanic Pollution and Climate: The 1783 Laki Eruption', *EOS* 63(32), 601–2 (1982).

279 *more than a hundred thousand flights*: Eurocontrol. 'Ash-cloud of April and May 2010: Impact on Air Traffic.' Retrieved from Eurocontrol website: https://www.eurocontrol.int/sites/default/files/article/attachments/201004-ash-impact-on-traffic.pdf

279 *99.99 per cent of its electricity from renewable sources*: 'ENERGY STATISTICS IN ICELAND 2019.' Retrieved from the website of Orkustofnun, Iceland's National Energy Authority: https://orkustofnun.is/gogn/os-onnur-rit/Orkutolur-2019-enska.pdf

281 *prompting critics to warn*: For example, Y. Zhou, 'Carbon capture and storage: A lot of eggs in a potentially leaky basket', International Council on Clean Transportation (17 January 2020).

281 Science *published the results*: J. Matter et al., 'Rapid carbon mineralization for permanent disposal of anthropogenic carbon dioxide emissions', *Science* 352(6291), 1312–14 (2016).

283 *an exhaustive 100-page report*: R. Socolow et al., 'Direct Air Capture of CO2 with Chemicals: A Technology Assessment for the APS Panel on Public Affairs', The American Physical Society (2011).

284 *When the filter is saturated*: V. Gutknecht et al., 'Creating a carbon dioxide removal solution by combining rapid mineralization of CO_2 with direct air capture', *Energy Procedia* 146, 129–34 (2018).

284 *Keith's system*: D. Keith et al., 'A Process for Capturing CO2 from the Atmosphere', *Joule* 2(8), 1573–94 (2018).

284 *carbon footprint of a single US family*: US carbon emissions amount to 15.5 tonnes per resident, according to the most recent World Bank data. Retrieved from https://data.worldbank.org/indicator/EN.ATM.CO2E.PC

284 *amount to well over 30 billion tonnes*: International Energy Agency. 'Global Energy Review: CO_2 Emissions in 2020.' Retrieved from https://www.iea.org/articles/global-energy-review-co2-emissions-in-2020

284 *nearly 100 million cars a year*: European Automobile Manufacturers Association. 'World Production.' Retrieved from https://www.acea.be/statistics/tag/category/world-production#:~:text=92.8%20million%20motor%20vehicles%20were%20produced%20globally%20in%202019 precise wording?

285 *taken by China to build wind and hydroelectric plants*: Data Retrieved from website of CDM Pipeline, run by UNEP DTU Partnership Centre on Energy, Climate and Sustainable Development: https://www.cdmpipeline.org/

285 *which, critics argued, it would have built anyway*: For example, Friends of the Earth, 'Trading in fake carbon credits: Problems with the Clean Development Mechanism.' Retrieved from https://foe.org/blog/2008-10-trading-in-fake-carbon-credits-problems-with-the-cle/

285 *It was often hard to gauge the impact of such schemes*: For an analysis of some of the questions surrounding tree planting offset schemes, see W. Anderegg et al., 'Climate-driven risks to the climate mitigation potential of forests', *Science* 368(6497), eaaz7005 (2020).

285 *losing a tenth of their area since 2000*: Data retrieved from the website of Global Forest Watch: https://www.globalforestwatch.org/dashboards/global/

285 *Even if you could somehow create and preserve*: Estimates of the

carbon sequestration potential of afforestation/reforestation vary widely depending on biome. I have used here an estimate given in an article co-authored by forty-nine international academics in *Science*, in October 2019, which suggested potential carbon sequestration of roughly 42 gigatonnes for a canopy cover restoration area of 900 million hectares. The latest UNEP Emissions Gap report estimated annual greenhouse gas emissions at 59.1 gigatonnes of carbon dioxide equivalent. A tonne of carbon dioxide contains 0.27 tonnes of carbon. Sources: J. Veldman et al., 'Comment on "The global tree restoration potential."' *Science* 366(6463) (2019); *UNEP Emissions Gap Report 2020*, accessed at https://www.unep.org/emissions-gap-report-2020; World Bank land area data accessed at https://data.worldbank.org/indicator/AG.LND.TOTL.K2

285 *a landmark report from the IPCC*: *Global Warming of 1.5°C. An IPCC Special Report.* IPCC (2018).

286 *including Bill Gates*: R. Sigurdardottir and A. Rathi, 'The Icelandic Startup Bill Gates Uses to Turn Carbon Dioxide Into Stone', Bloomberg (5 March 2021).

287 *Adolf Hitler used it . . .*: D. Leckel, 'Diesel Production from Fischer–Tropsch: The Past, the Present, and New Concepts', *Energy Fuels* 23 (5), 2342–58 (2009).

288 *One kilogram of jet fuel provides*: J. Holladay et al. 'Sustainable Aviation Fuel: Review of Technical Pathways.' Report published by the US Department of Energy at https://www.energy.gov/sites/prod/files/2020/09/f78/beto-sust-aviation-fuel-sep-2020.pdf

288 *try to imagine a Boeing 767*: I've used the technical details for the 767-400 Extended Range, which weighs 150 tonnes without fuel, and has a fuel capacity of 91,377 litres. Jet fuel has a density of about 0.8kg/l. 767-400 technical details retrieved from Boeing's website at https://www.boeing.com/commercial/aeromagazine/aero_03/textonly/ps01txt.html. Jet fuel density data from Chevron, 'Aviation Fuels Technical Review.' Retrieved from https://www.chevron.com/-/media/chevron/operations/documents/aviation-tech-review.pdf

289 *found his path blocked*: P. Smyth, 'Frans Timmermans not nominated for fear of east-west divisions', *Irish Times*, 30 June 2019.

290 *a major financier of the Carbfix/Climeworks collaboration*: Factsheet on Horizon 2020 grant, Grant agreement ID: 764760. Retrieved

from European Commission website at https://cordis.europa.eu/project/id/764760

290 *the level that leading analysts reckon is needed*: See, for example, 'Effective Carbon Rates 2021: Pricing Carbon Emissions through Taxes and Emissions Trading.' OECD report retrieved from https://www.oecd.org/tax/tax-policy/effective-carbon-rates-2021-0e8e24f5-en.htm

CHAPTER 20

295 *'Building an ecological civilisation is vital . . .*: 'Full text of Xi Jinping's report at 19th CPC National Congress', Xinhua (4 November 2017)

297 *65 gigawatt-hours of batteries*: Communication from BYD spokesperson.

297 *six billion iPhones*: The iPhone 12 has a battery capacity of 10.78 Wh, according to Phone Arena: https://www.phonearena.com/news/apple-iphone-12-pro-max-mini-vs-11-battery-capacity_id127955

297 *Wang Chuanfu, a former government metallurgical researcher*: M. Campbell and Y. Tian, 'The World's Biggest Electric Vehicle Company Looks Nothing Like Tesla', Bloomberg (16 April 2019).

298 *16,000 BYD buses and 22,000 of its e6 taxis*: Communication from BYD spokesperson.

298 *in the Chinese city of Yinchuan*: 'BYD-built monorail opens in Yinchuan', *Metro Report International*, 5 September 2017.

298 *in the Brazilian port city of Salvador*: 'Salvador monorail construction starts', *Metro Report International*, 4 March 2020.

298 *Shenzhen, which in 1977*: E. Vogel, *Deng Xiaoping and the Transformation of China,* Chapter 7. Harvard University Press (2011).

298 *average income was nearly twenty times higher*: According to World Bank data, 1977 per capita income (in current US$) was $185 in China and $3,429 in Hong Kong. Data Retrieved from https://data.worldbank.org/indicator/NY.GDP.PCAP.CD?locations=HK

299 *megacity of 13.4 million*: 'Shenzhen Basics.' Retrieved from website of Shenzhen government: http://www.sz.gov.cn/en_szgov/aboutsz/profile/content/post_1357629.html

299 *annual export revenue exceeding that of Brazil*: Shenzhen's exports

in 2019 were worth CNY1.67 trillion, equivalent to $242 billion at the average exchange rate for that year (according to Macrotrends data). Brazil's exports in 2019 were worth $224 billion. Sources: 'SZ exports rank No. 1 for 27 straight years.' Retrieved from website of Shenzhen government: http://www.sz.gov.cn/en_szgov/news/latest/content/post_7871613.html; J. McGeever and M. Ayres, 'Brazil trade surplus shrinks 20% in 2019 to its smallest in four years', Reuters (2 January 2020).

299 *began burning more coal than the rest of the world combined*: Since 2011, according to the CSIS China Power Project. 'How Is China's Energy Footprint Changing?'. Retrieved from https://chinapower.csis.org/energy-footprint/

299 *at a rate of about one per week*: 'China Dominates 2020 Coal Plant Development.' Briefing paper from Global Energy Monitor and the Center for Research on Energy and Clean Air (February 2021).

299 *the highest volume of any country*: 'Factbox: China becomes the world's No. 1 auto market.' Reuters (8 January 2010).

300 *a change to the Chinese constitution*: Deng, Y., 'Amendments reflect CPC's resolve', *China Daily*, 15 November 2012.

300 *pollution in Beijing's notoriously smoggy skies*: L. Lim, 'Beijing's "Airpocalypse" Spurs Pollution Controls, Public Pressure', NPR (14 January 2013).

300 *the government had nurtured*: C. Gang, 'China's Solar PV Manufacturing and Subsidies from the Perspective of State Capitalism', *Copenhagen Journal of Asian Studies* 33(1), 90–106 (2015).

301 *China built more new solar capacity*: According to Chinese government data, China installed 125GW of solar PV capacity in the period 2013–2017. According to the IEA-PVPS programme, about 96.5GW of capacity had been installed globally at the end of 2012. Sources: Annual solar photovoltaic statistics retrieved from the website of China's National Energy Administration at http://www.nea.gov.cn/; 'A Snapshot of Global PV 1992–2012.' Report by IEA-PVPS, retrieved from https://iea-pvps.org/wp-content/uploads/2020/01/PVPS_report_-_A_Snapshot_of_Global_PV_-_1992-2012_-_FINAL_4.pdf

301 *a $15 billion deficit*: L. Hook and L. Hornby, 'China's solar desire dims', *Financial Times*, 8 June 2018.

301 *suddenly declaring that*: '2018年光伏发电有关事项的通知.' Statement from China's National Development and Reform Commission,

retrieved from https://www.ndrc.gov.cn/xxgk/zcfb/tz/201806/t20180601_962736.html

301 *sparking fears for the survival . . .*: G. Luo et al., 'China's Solar Industry Dims After Subsidy Cuts', *Caixin* Global, 24 June 2018.

302 *sending the cost of solar power crashing more than 80 per cent in the past decade*: 'Renewable Power Generation Costs in 2019', International Renewable Energy Agency (2020).

302 *It's still building . . . and the rest of the world . . .*: 'World Adds Record New Renewable Energy Capacity in 2020.' Press release by International Renewable Energy Agency (5 April 2021).

303 *Elon Musk tosses his jacket aside*: Video posted by Bloomberg Markets and Finance at https://www.youtube.com/watch?v=ZucUZZkIQw0

303 *Depending on the configuration*: 'Xpeng P7 Launches in China.' Press release by Xpeng Motors (27 April 2020).

303 *more than 1,000 Chinese people*: Ponciano, J., 'The Countries With The Most Billionaires 2020', *Forbes*, 8 April 2020.

303 *more than 1,000 Chinese people*: 'Hurun Global Rich List 2021', Hurun Report (2 March 2021).

303 *China's biggest ever internet buyout*: P. Carsten, 'Alibaba to buy out UCWeb in China's biggest internet merger', Reuters (11 June 2014).

304 *sales of $25 billion*: T. Hsu, 'Alibaba's Singles Day Sales Hit New Record of $25.3 Billion', *New York Times*, 10 November 2017.

304 *slashed subsidies for electric cars by half*: A. Kharpal, 'As China cuts support for its electric carmakers, auto firms could face a "war of attrition"', CNBC (19 June 2019).

304 *Xpeng's main rival Nio . . .*: C. Trudell, 'NIO Reaches Government Deal Bernstein Calls a Bailout', Bloomberg (25 February 2020).

305 *enough to power several thousand Chinese homes*: For this calculation I have used an estimate for urban Chinese household annual electricity consumption of 1,690kWh, as reported in a 2017 paper by researchers at Renmin University: S. Hu et al., 'A survey on energy consumption and energy usage behavior of households and residential building in urban China', *Energy and Buildings* 148, 366–78 (2017).

306 *top four global wind turbine producers*: 'GWEC releases Global Wind Turbine Supplier Ranking for 2020.' Press release by Global Wind Energy Council (23 March 2021).

307 *one of President Xi's most ambitious ideas*: Downie, E., 'China's Vision for a Global Grid', *Reconnecting Asia* (13 February 2019).

307 *suspicions of forced labour*: D. Murtaugh et al., 'Secrecy and Abuse Claims Haunt China's Solar Factories in Xinjiang', Bloomberg (13 April 2021).

CHAPTER 21

309 *'We can, and we will, deal with climate change . . .*: Transcript of Joe Biden's speech retrieved from CNBC website: https://www.cnbc.com/2020/08/21/joe-biden-dnc-speech-transcript.html
311 *home to oil wells since 1901*: A. Johnson, 'The Early Texas Oil Industry: Pipelines and the Birth of an Integrated Oil Industry, 1901–1911', *Journal of Southern History* 32(4), 516–28 (1966).
311 *US oil prices briefly turned negative*: D. Sheppard et al., 'US oil price below zero for first time in history', *Financial Times*, 21 April 2020.
311 *the break-even point for many shale producers*: M. Passwaters, 'Half of producing shale oil wells are profitable at $40/bbl, analyst says', S&P Global Market Intelligence (21 August 2020).
311 *slashed their long-term forecasts*: R. Katakey, 'BP Says the Era of Oil-Demand Growth Is Over', Bloomberg (14 September 2020).
312 *Lithium-ion batteries still look prohibitively expensive*: W. Cole and A. W. Frazier, 'Cost Projections for Utility-Scale Battery Storage', National Renewable Energy Laboratory. NREL/TP-6A20-73222 (2019).
314 *According to a lawsuit later filed by Awan's widow*: Complaint by Liliana Awan vs. Tesla Motors Inc. and Tesla Florida Inc. in the Circuit Court of the 17th Judicial Circuit. Case Number: CACE-19-021110 Division: 08 (filed 10 October 2019).
315 *For decades, scientists have known*: K. Takada, 'Progress and prospective of solid-state lithium batteries', *Acta Materialia* 61(3), 759–70 (2013).
315 *Silicon Valley's newest startup billionaire*: I met Tim Holme on 10 December 2020. The previous day, Quantumscape's share price had risen 30 per cent to close at $75.14. This valued the company at $27.27 billion, and Holme's 4.06 per cent stake at $1.11 billion. Sources: S&P Capital IQ database; Form S-1 Registration Statement filed with the Securities & Exchange Commission by QuantumScape Corporation (17 December 2020): https://sec.report/Document/0001193125-20-320220/
315 *a market valuation of $48 billion*: On 22 December 2020

Quantumscape's share price recorded an intraday high of $132.73, valuing the company at $48.17 billion. Ford Motor Co. shares recorded an intraday high on the same day of $11.64, valuing Ford at $35.06 billion. Sources: S&P Capital IQ database; https://ir.quantumscape.com/stock-info/default.aspx; https://shareholder.ford.com/investors/stock-information/default.aspx

316 *The undoing of previous efforts in this field had been dendrites*: H. Liu et al., 'Controlling Dendrite Growth in Solid-State Electrolytes', *ACS Energy Lett.* 5(3), 833–43 (2020).

316 *a short-selling attack by a hedge fund*: A. Rathi, 'QuantumScape Defends Its Battery Breakthrough Against the Short Sellers', Bloomberg (27 April 2021).

317 *plans to have a factory producing its electric car batteries from 2024*: Author interview with QuantumScape CEO Jagdeep Singh (3 December 2020).

318 *KiOR, a Mississippi outfit*: K. Fehrenbacher, 'A Biofuel Dream Gone Bad', *Fortune*, 4 December 2015.

318 *Nordic WindPower, an offshoot from a Swedish government project*: P. Koepp, 'Turbine-maker Nordic Windpower files for liquidation bankruptcy', *Kansas City Business Journal*, 19 October 2012.

318 *the doomed Infinia*: 'Infinia, maker of distributed power engines, seeks bankruptcy protection.' Press release from Infinia Corporation (26 September 2013).

319 *publicly mocking his rivals' attachment to golf and yachting*: J. Cook, 'Vinod Khosla on failure, thinking big and why the best entrepreneurs are under 25', Geekwire (30 September 2011).

319 *he went to the US Supreme Court*: N. Bowles, 'Every Generation Gets the Beach Villain It Deserves', *New York Times*, 30 August 2018.

319 *a virtual* Who's Who: K. Dolan, 'Bill Gates, Mark Zuckerberg & More Than 20 Other Billionaires Launch Coalition To Invest In Clean Energy', *Forbes*, 29 November 2015.

320 *Range Fuels went bust in 2012*: Herndon, A., 'Range Fuels Sells Government-Backed Biofuel Plant to LanzaTech', Bloomberg (4 January 2012).

321 *the world's busiest airport*: 'ACI reveals top 20 airports for passenger traffic, cargo, and aircraft movements.' Press release from Airports Council International (19 May 2020).

322 *Virgin Atlantic flew a plane*: G. Topham, 'First commercial flight partly fuelled by recycled waste lands in UK', *Guardian*, 3 October 2018.

322 *China's Shougang started churning out*: 'Commercial CCU Plant using LanzaTech Technology Receives RSB Advanced Products Certification.' Press release from LanzaTech (28 January 2021).

324 *they would harness the force of nuclear fusion*: 'Joint Soviet-United States Statement on the Summit Meeting in Geneva' (21 November 1985). Retrieved from the website of the Ronald Reagan Presidential Library & Museum: https://www.reaganlibrary.gov/archives/speech/joint-soviet-united-states-statement-summit-meeting-geneva

324 *they began preparations*: 'The Iter Story.' Retrieved from the website of ITER: https://www.iter.org/proj/ITERHistory

324 *estimated by the US Government at $65 billion*: D. Kramer, 'ITER disputes DOE's cost estimate of fusion project', *Physics Today*, 16 April 2018.

324 *generating energy for the first time in 2035*: 'What Will Iter Do?' Retrieved from the website of ITER: https://www.iter.org/proj/Goals

325 *Per kilogram, a quietly rotting pile of compost generates far more heat*: 'Sun Fact Sheet.' Retrieved from NASA website at https://nssdc.gsfc.nasa.gov/planetary/factsheet/sunfact.html; E. Harper et al., 'Physical management and interpretation of an environmentally controlled composting ecosystem', *Australian Journal of Experimental Agriculture* 32(5) 657–67 (1992).

325 *plasma that reached 100,000,000°C*: 'Alcator C-Mod tokamak.' Retrieved from website of Plasma Science and Fusion Center, Massachusetts Institute of Technology: https://www.psfc.mit.edu/research/topics/alcator-c-mod-tokamak

325 *six times hotter than the sun's core*: 'Anatomy of our Sun.' Retrieved from the website of the European Space Agency: http://www.esa.int/ESA_Multimedia/Images/2019/10/Anatomy_of_our_Sun

326 *ITER's reactor in France will use 10,000 tonnes of magnets* – 'Magnets.' Retrieved from the website of ITER: https://www.iter.org/mach/Magnets

326 *together willing to bet almost $200 million*: Data Retrieved from Crunchbase: https://www.crunchbase.com/organization/commonwealth-fusion-systems/company_financials

326 *Bob's superconducting magnet, which will be . . .*: 'HTS Magnets.' Retrieved from the website of Commonwealth Fusion Systems: https://cfs.energy/technology/hts-magnets; 'Magnets.' Retrieved from the website of ITER: https://www.iter.org/mach/Magnets

CHAPTER 22

329 *'With a climate crisis looming . . .*: 'Amnesty challenges industry leaders to clean up their batteries.' Press release from Amnesty International (21 March 2019).

331 *more cobalt than a thousand iPhones*: The battery of the first Tesla Model S contained 11 kilograms of cobalt, according to Benchmark Mineral Intelligence. According to an analysis at a metallurgical lab commissioned by journalist Brian Merchant, the iPhone 6 contained 6.6 grams of cobalt. Sources: 'Panasonic Reduces Tesla's Cobalt Consumption By 60% In 6 Years.' Benchmark Intelligence Blog (3 May 2018); B. Merchant, 'Everything That's Inside Your iPhone', *Vice*, 15 August 2017.

331 *most of the world's known reserves were in Congo*: US Geological Survey. 'Cobalt Data Sheet - Mineral Commodity Summaries 2020.' Retrieved from https://pubs.usgs.gov/periodicals/mcs2020/mcs2020-cobalt.pdf

334 *When King Leopold . . .*: See D. Van Reybrouck, *Congo: The Epic History of a People*, Ecco (2014); Hochschild, A., *King Leopold's Ghost*, Picador (2019).

334 *'I don't want us to miss an opportunity to get ourselves a piece . . .'*: R. Cornet, 'A propos de l'ouvrage du baron Pierre van Zuylen: "L'échiquier congolais ou le secret du Roi"', *Académie Royale des Science Coloniales*, Bulletin des Séances V-1959-4, 857-863 (1959).

334 *BMW announced . . .*: W. Clowes, 'BMW to Source Cobalt Directly From Australia, Morocco Mines', Bloomberg (24 April 2019).

334 *Elon Musk tweeted*: Tweet by @elonmusk: 'We use less than 3% cobalt in our batteries & will use none in next gen' (13 June 2018).

335 *one of the largest and poorest in the world*: World Bank Data retrieved at https://data.worldbank.org/indicator/AG.LND.TOTL.K2 and https://data.worldbank.org/indicator/AG.LND.TOTL.K2

335 *the increasingly dysfunctional and kleptocratic thirty-year rule of Mobutu Sese Seko*: See Van Reybrouck, *Congo;* M. Wrong, *In the Footsteps of Mr Kurtz*, Fourth Estate (2012).

335 *he was appointed interior minister*: 'Congo (Democratic Republic): The prime minister forms a new government', Economist Intelligence Unit (1 June 2012).

335 *a country the size of . . .*: World Bank data retrieved from https://data.worldbank.org/indicator/AG.LND.TOTL.K2?most_recent_value_desc=true

335 *miles of paved road*: Sources: 'Field Listings – Roadways.' CIA World Factbook, retrieved 9 May 2021 at https://www.cia.gov/the-world-factbook/field/roadways/; 'Road lengths and conditions, 2018–19: revised.' Welsh Government Statistical Bulletin (22 August 2019).

336 *a lawsuit will be filed by US activists*: H. Dempsey, 'Tech giants sued over child deaths in DRC cobalt mining', *Financial Times*, 16 December 2019.

337 *many thousands of others die each year*: S. Cha et al., 'Effects of improved sanitation on diarrheal reduction for children under five in Idiofa, DR Congo: a cluster randomized trial', *Infectious Diseases of Poverty* 6,137 (2017).

337 *a corruptly mismanaged economy*: See T. Burgis, *The Looting Machine: Warlords, Tycoons, Smugglers and the Systematic Theft of Africa's Wealth*, Chapter 2, William Collins (2015); 'Regime Cash Machine.' Report by Global Witness (July 2017); Transparency International Corruption Perceptions Index 2020.

337 *has the lowest per capita health spending in the world*: World Bank Data for 2018, accessed in May 2021 at https://data.worldbank.org/indicator/SH.XPD.CHEX.PC.CD?most_recent_value_desc=false

AFTERWORD

345 *at most a few million*: H. Leridon, 'Human populations and climate: Lessons from the past and future scenarios', *Comptes Rendus Geoscience* 340(9–10), 663–9 (2008).

345 *one-tenth the speed*: E. Jansen et al., 'Palaeoclimate', in *Climate Change 2007: The Physical Science Basis. Contribution of Working Group I to the Fourth Assessment Report of the Intergovernmental Panel on Climate Change*, Cambridge University Press (2007), p. 451.

345 *Most of the migration will take place within developing countries*: *Groundswell: Preparing for Internal Climate Migration.* World Bank report (March 2018).

346 *worst extended drought on record*: C. Kelley et al., 'Climate change in the Fertile Crescent and implications of the recent Syrian drought', *PNAS* 112(11), 3241–6 (2015).

346 *For each of its inhabitants, the United States emits . . .*: World Bank data retrieved in May 2021 at https://data.worldbank.org/indicator/EN.ATM.CO2E.PC

346 *Even China . . .*: Our World in Data calculations, retrieved in May 2021 from https://ourworldindata.org/co2-emissions

346 *All but a handful of rich countries are failing . . .*: 'Aid by DAC members increases in 2019 with more aid to the poorest countries.' OECD press release (16 April 2020).

346 *cited the global crisis*: G. Parker and J. Cameron-Chileshe, 'Minister quits as chancellor comes under attack over foreign aid cuts', *Financial Times,* 25 November 2020.

347 *opinion polls showed overwhelming approval*: C. Hirsch, 'European citizens support strict coronavirus lockdown, say polls', *Politico*, 3 April 2020.